Answe
Revision Q
for High

Physics

Lyn Robinson

Principal Teacher of Physics

Williamwood High School, Clarkston

Published by
Chemcord
Inch Keith
East Kilbride
Glasgow

ISBN 1 870570 77 4

Printed by Bell and Bain, Glasgow

Contents

KINEMATICS

Exercise 1.1 Vectors and Scalars

1. a) A vector quantity has magnitude and direction.
b) A scalar quantity has magnitude only.

2. a) *Vectors*: weight, momentum, velocity, acceleration
Scalars: mass, kinetic energy
b) *Vectors*: force, momentum, weight
Scalars: mass, distance
c) *Vectors*: velocity, displacement
Scalars: power, speed, work, potential energy
d) *Vectors*: pressure, force, impulse
Scalars: power, time, temperature

3. a) False
b) True
c) False
d) True
e) False

4. a) 50 km, 36.9° E of N
b) 35 km h^{-1}
c) 25 km h^{-1} at 36.9° E of N

5. a) 5 km E
b) 12 km

6. a) 1000 m 36.9° W of N
b) 4 m s^{-1}
c) 1.67 m s^{-1} 36.9° W of N

7. a) 1422 m 39° W of N
b) 39° E of S

8. 6.4 m s^{-1} 51.3° E of N

9. 5.2 m s^{-1} 22.6° E of N

10. 3.7 m s^{-1} N

11. a) 17.3 m s^{-1}
b) 10 m s^{-1}

12. a) 38.3 m s^{-1}
b) 32.1 m s^{-1}

13. a) 100 N
b) 173 N

14.

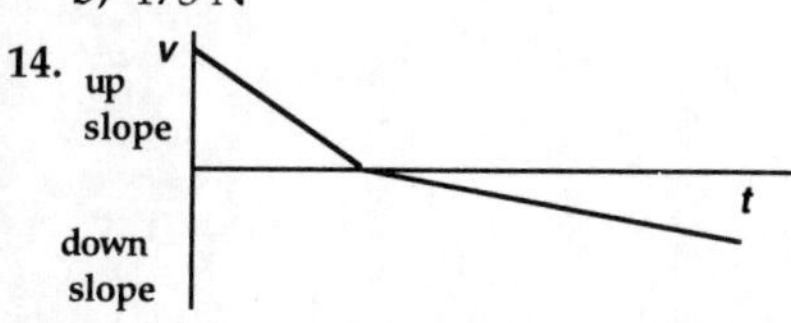

Exercise 1.2 Velocity and Acceleration Graphs

1. a) a/m s^{-2}; -5; 2; 4; t /s

b) a/m s^{-2}; 10; 2; 7; 8; t /s; -20

c) a/m s^{-2}; 2; -4; 2; 4; 6; 8; t /s

2. 40 m

3. a/m s^{-2}; 3; 2; 1; -1; -2; 2; 4; 6; t /s

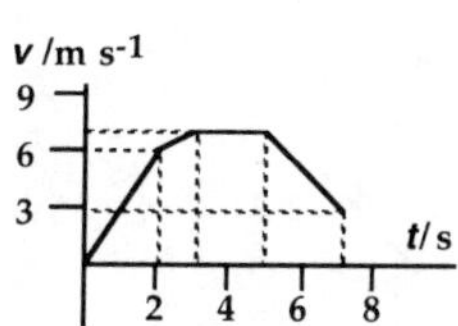

4. a) 625 m; 125 m; ~~3.13 m s^{-1}~~ 0.052 ms^{-1}
b) 2500 m; -500 m; -1.67 m s^{-1}

5. a) acceleration
b) displacement
c) velocity

6. 12.5 m s^{-1}

7. a) 16 m s^{-1}
b) 14 m s^{-1}

8. velocity; time

9. Speed

10. **OP** increasing acceleration
PQ constant acceleration
QR constant velocity
RS decreasing acceleration
ST constant acceleration
(lower than **PQ**)

11. a)

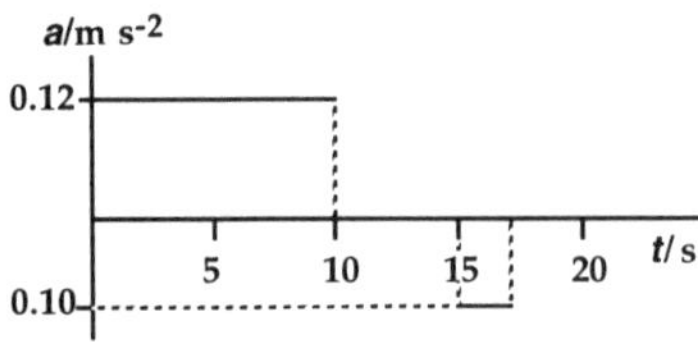

b)

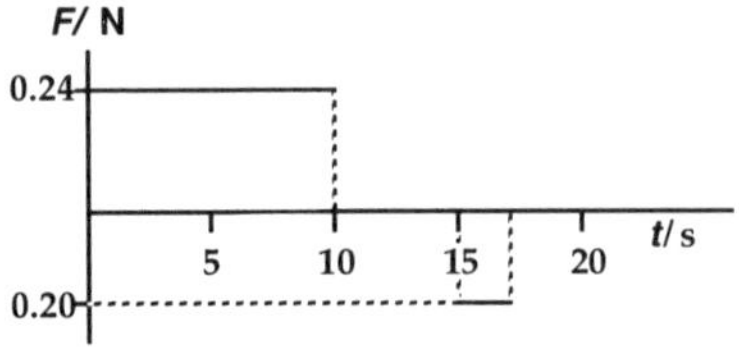

c) Trolley accelerating down a uniform slope.
d) Trolley begins to go uphill.
e) 20.2 m
f) 0.88 m s^{-1}

12. a)

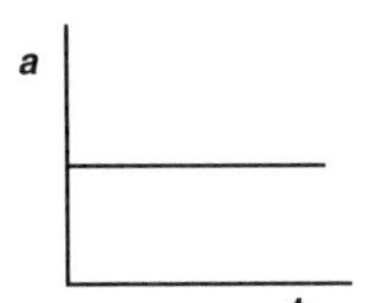

b)

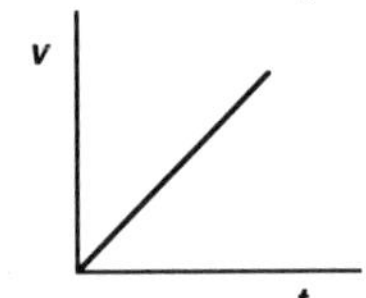

c)

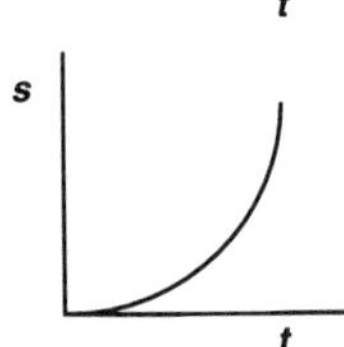

13. 20 000 kg

14. a) **A** - engine cuts off
B - reached top of flight
C - crashes
b) 900 m
c) 3.3 m s^{-2}
d) 130 m above
e)

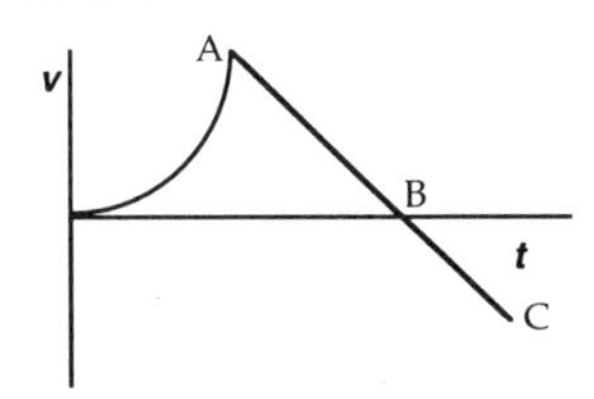

15. a)

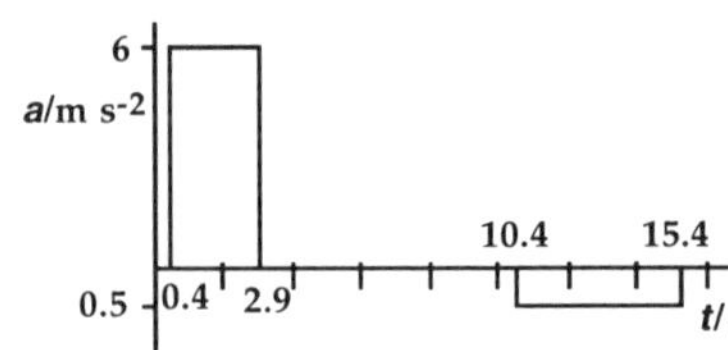

b)

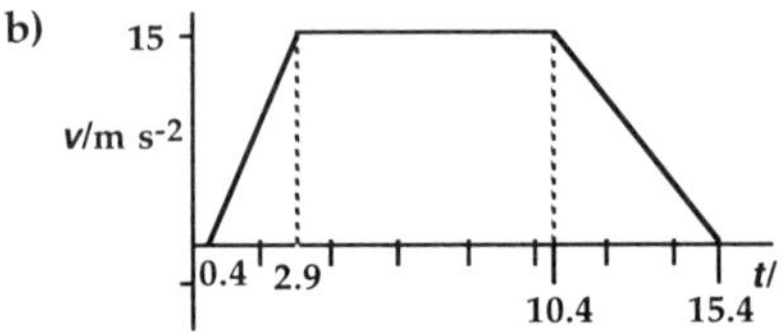

c) 14.3 s
d) 15.4 s
e) No, as hare reaches 200 m first.

16. a) **AB** - accelerating down under gravity
BC - comes to stop on the ground
CD - rapidly accelerating in an upward direction, leaving the ground again at **D**
b) 4 bounces.
c) Kinetic energy is lost as speed decreases at each bounce.

Exercise 1.3 Equations of Motion

1. 26 m s^{-1}
2. 44.1 m
3. 2000 m s^{-2}
4. -9 m s^{-2}
5. 3.3 m s^{-2}
6. 500 m
7. 100 m
8. -2.67×10^{-3} m s^{-2}
9. 0.505 s
10. 1.9 m s^{-1}
11. 273 m s^{-1}
12. 7.8 s
13. 1.52 s
14. 3.06 s
15. Distance travelled if continued at constant speed.

16. a)

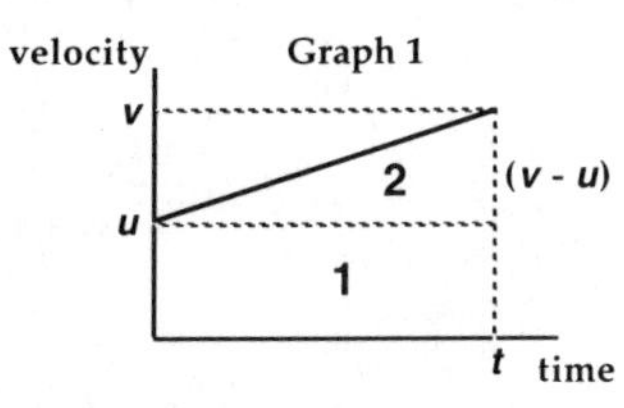

s = area 1 + area 2

$s = ut + \frac{1}{2}(v - u)t$

but $v = u + at \Rightarrow v - u = at$

$s = ut + \frac{1}{2}(at)t$

$s = ut + \frac{1}{2}at^2$

b)

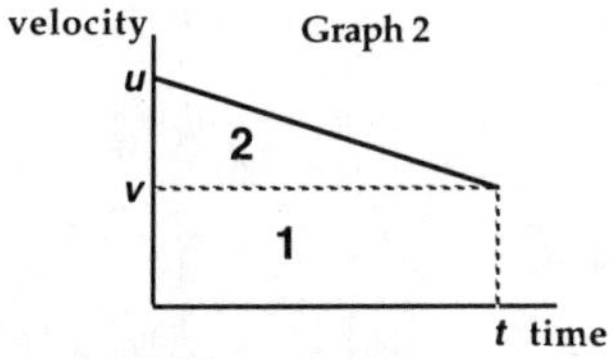

s = area 1 + area 2

$s = \frac{1}{2}(v - u)t + ut$

$s = \frac{1}{2}vt - \frac{1}{2}ut + ut$

$s = \frac{1}{2}vt + \frac{1}{2}ut$

$s = \frac{(u + v)}{2}t$

17. 147.5 m
18. 97.5 m
19. 1360 m
20. 38.1 m
21. D
22. C

Exercise 1.4 Projectiles - Horizontal Projection

1. 50 m s^{-1}, 36.9° to horiz.

2. a) 7.07 s

b)

v_h/m s^{-1}

180

7.07 t/s

c)

v_v/m s^{-1}

69.3

7.07 t/s

3. 20 m s^{-1}

4. 3.06 s

5. a)

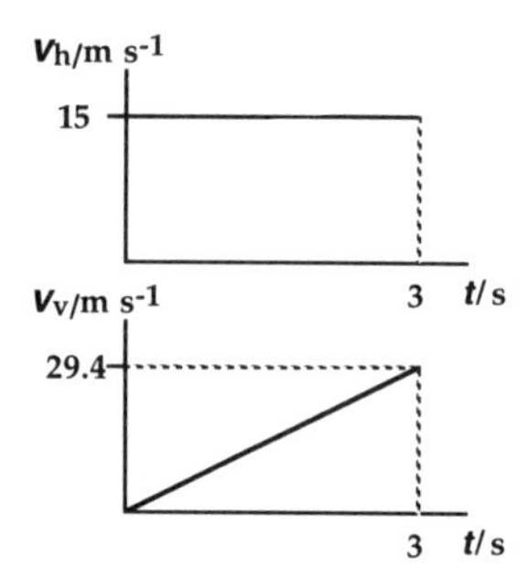

b) 24.7 m s^{-1} 37.4° to horizontal; assuming no air resistance

c) E_p -> E_k

6. 50 m

7. a) 0.606 s

b) 9.9 m s^{-1}

8. a) 3.03 s

b) 29.7 m s^{-1}

9. 19.8 m s^{-1}

10. 7.83 m s^{-1}

11. a) 3.96 m s^{-1}

b) 0.505 s

c) 2 m

Exercise 1.5 Projectiles - Oblique Projection

1. a) 150 m s^{-1}
b) 54.7 m s^{-1}
c) 5.58 s
d) 11.2 s
e) 1680 m

2. a) 25.4 m s^{-1}
b) 2.36 s

3. a) 14.7 m s^{-1}
b) 8.54 m s^{-1}

4. a) 36.3 m s^{-1}
b) 16.9 m s^{-1}
c) 131 J; top of flight
d) i) 9.8 m s^{-2} down
ii) 0
e) 1.72 s
f) 3.45 s
g) 125 m
h)

v_h/m s^{-1}
36.3
3.45 t/s

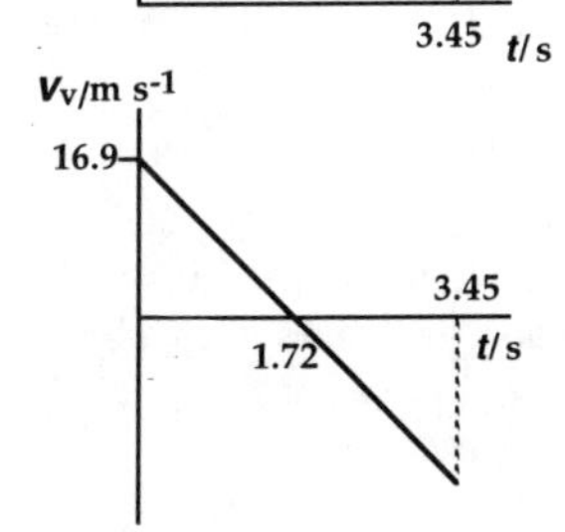

5. 45.9 m

6. a) 240 m
b) 15 m
c) 50 m s^{-1} 36.9^o to horizontal

7. 2.86 s

8. a) 30°
b) i) same
ii) same
iii) same
c) at the start

9. 1.06 m above ground, hits

10. a) 12 m s^{-1}
b) 312 m
c) 1.6 m s^{-2}

11. a) 13.6 m s^{-1}
b) 7.84 m s^{-1}
d) 5.5 m

12. a) 20.78 m s^{-1}
b) 12 m s^{-1}
c) 2.45 s
d) 12

DYNAMICS

Exercise 2.1 Forces

1. a) 1000 N
   b) 1960 N
   c) 1.875 N
   d) 5390 N

2. 500 N

3. 12 m s^{-2}

4.

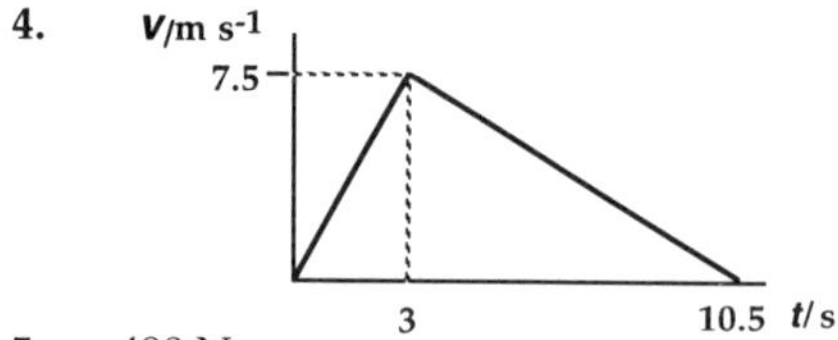

5. 490 N

6. 54.9 N

7. a)

Thrust
240 000 N

Weight
98 000 N

   b) 14.2 m s^{-2}
   c) Acceleration increases because mass decreases as fuel burnt.

8. 26 400 N

9. a)

Buoyancy force

Weight Tension

   b) 5650 N
   c) 750 N

10. a)

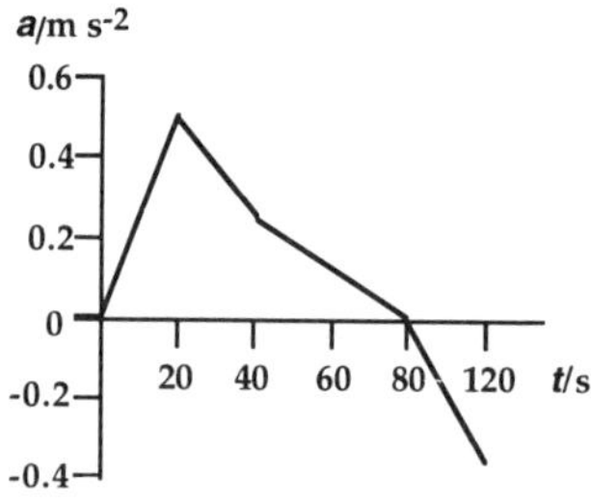

   b)

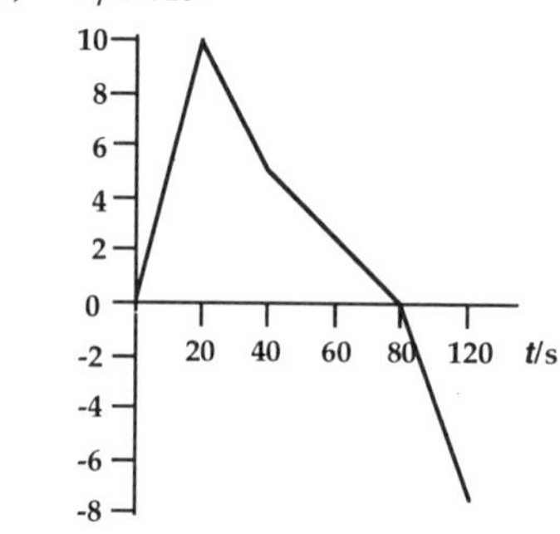

11. a) 1 m s^{-2}
    b) 15

12. a) 8.44×10^6 N
    b) 3.84 m s^{-2}
    c) i) Acceleration increases.
       ii) Mass decreases as fuel burnt.

13. a) Force due to Moon's gravity on craft
    = 1.6 x 15 000 N
    = 24 kN
    which is equal and opposite to the rocket thrust, hence the speed will remain constant.
    b) i) -0.1 m s^{-2}
       ii) 0.2 m s^{-1}
       iii) 19.8 m

Exercise 2.2 Internal Forces

1. 14 N
2. a) 6 N
   b) i) Equal and opposite, same mass, same force in opposite direction
      ii) -4 N
3. 0
4. 3000 N
5. a) 49 N
   b) 0
6. a) Accelerates at 4 m s^{-2}
   b) **A** - accelerates at 12 m s^{-2} from 12 m s^{-1}
      B - moving at constant velocity of 12 m s^{-1}
7. 1.5 ***d***

Exercise 2.3 Components of Force

1. a) 13 N
   b) 7.5 N
2. 80 N
3. 390 N
4. 1.73×10^3 N
5. a) 1000 N
   b) The tension will be less.
6. 11.3 N

Exercise 2.4 Force on a Slope

1. 9.8 N
2. a) 41.4 N
   b) 159 N
3. 216 N
4. a) 14.7 N
   b) 4.9 m s^{-2}
   c) 9.8 m
5. a) i) 137 N
      ii) 2.87 m s^{-2}
   b) No effect, provided slope constant.
   c) Friction increases as the velocity increases until balances component of weight.

Exercise 2.5 Lifts

1. a) 1.3 m s^{-2}
   b) 3 s
2. 11.8 N
3. 0.348 N
4. 0.245 m s^{-2} up
   -0.245 m s^{-2} down
5. Accelerating up or decelerating while lift moves down.
6. a) 9.8 N
   b) 9.8 N
   c) 12.8 N
   d) 7.8 N
   e) 13.8 N
7. a) R > F
   b) R = F
   c) R < F
   d) R < F
   e) R > F
8. a) The lift accelerates for the first two seconds, moves at a steady speed for the next six seconds and then decellerates for the final two seconds.
   b) 140 N↓; 0; 140 N↑
   c)

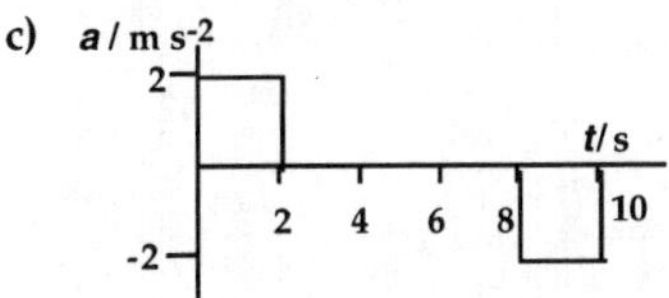

Exercise 2.6 Momentum

1. a) 11 900 kg m s^{-1}
   b) 1.53 kg m s^{-1}
   c) 1.26 kg m s^{-1}
   d) 1.2×10^7 kg m s^{-1}

2. -1 m s^{-1}

3. 350 m s^{-1}

4. 2 m s^{-1}

5. a) -0.25 m s^{-1}
   b) 0
   c) 4.75 m s^{-1}

6. a) 1 m s^{-1} right
   b) 8 m s^{-1} right

Exercise 2.7 Elastic and Inelastic Collisions

1. a) Momentum and E_k conserved
   b) Momentum only conserved

2. a) 0.2 m s^{-1} right
   b) inelastic

3. a) 3 m s^{-1} right
   b) inelastic

4. a) 3[illegible] m s^{-1} right
   b) elastic

5. 2 J

6. a) No
   b) 2 m s^{-1}
   c) halved

7. 400 J

Exercise 2.8 Impulse

1. $I = Ft$ (N s)
   $I = mv - mu$ (kg m s^{-1})

2. 6 N s

3. a) 2.4×10^4 kg m s^{-1}
   b) 4.8×10^5 N

4. 80 N

5. Impulse or change in momentum

6. a) 40 m s^{-1}
   b) 12 m s^{-1}
   c) 20 m s^{-1}

7.

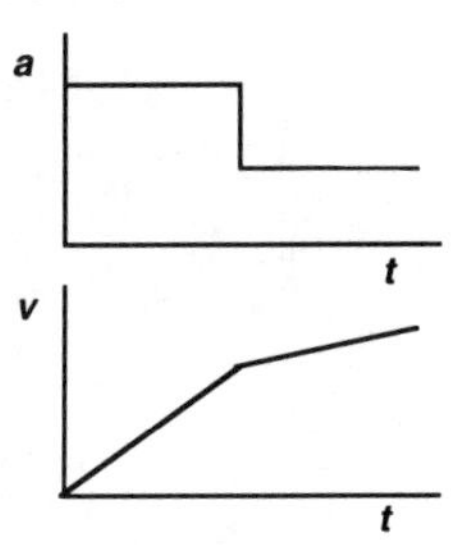

8. Skull takes longer to come to rest, so average force less for same change in momentum.

9. a) i)

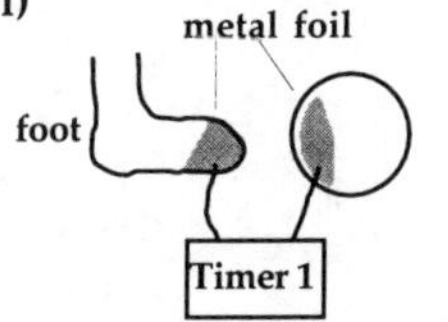

   ii) Kick ball through light gate and measure diameter of ball.

   iii) Time of contact, boot to ball, from timer 1 - t_1
   Time through light gate 2 - t_2
   $v = \frac{\text{diameter of ball}}{t_2}$

   b) $Ft = mv - mu$
   $u = 0$. Also need mass of ball.

10. 8 N

11. 30 m s^{-1}

12. a) 150 m^3
    b) 193.5 kg
    c) 9675 N

13. a) Calculate total area under the graph. Each box is 0.1 N s.
    b) 1.85 kg m s^{-1}
    c) 40.2 m s^{-1}

14. a) 0.55 m s^{-1}
    b) 0.2 kg m s^{-1}

15. a) Force of rocket pushing gas back = force of gas pushing rocket forward (Newton's 3rd Law)
    b) i) 4.5×10^6 kg m s^{-1}
    ii) 4.5×10^6 N
    iii) Mass decreases as fuel burnt.

16. a) i) 7.5 N s
    ii) 50 m s^{-1}
    b)

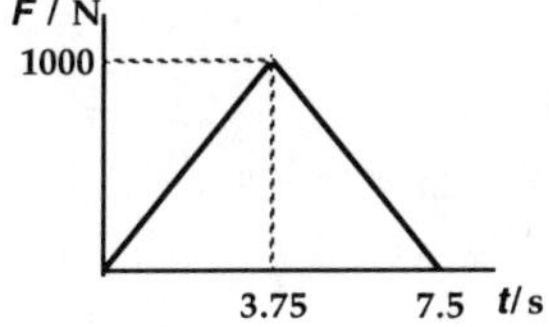

(same area, longer time, lower force)

    c) Mass increases, so velocity decreases.

Exercise 2.9 Energy

1. E_p -> E_k
   -> energy of deformation + E_h
   -> $(E_k)_{less}$
   -> $(E_p)_{less}$

2. 25 m s^{-1}

3. a) 800 J
   b) 600 J
   c) 200 J

4. a) 80 J
   b) 56.6 m s^{-1}
   c) All E_p stored in catapult becomes E_k of stone.

5. 1 : 4

6. 2.42 x 10^8 J

7. 60 J

8. a) i) 1.98 m s^{-1}
   ii) 3.96 m s^{-1}
   b) Mass

9. 40 m s^{-1}

10. a) 5 x 10^{-4} s
    b) 3.2 x 10^4 N

11. 17.9 m s^{-1}

12. a) 44.1 J
    b) 14.1 J
    c) 30 J
    d) 3.53 N

13. 0.051 m

14. a) 82 320 W
    b) Need extra power to accelerate initially **or** motor not 100% efficient.
    c) Lost as E_h in the brakes.
    d) $P = Fv$ but F decreases since $F = mg \sin \theta$, so P decreases

Exercise 2.10 Mechanics - Mixed problems

1. a) Calculate time for vertical displacement. t = 0.4 s

   Horizonal veleocity $= \frac{d}{t} = \frac{0.2}{0.4}$

   $= 0.5 \text{ m s}^{-1}$

   b) In the absence of an external force, the total momentum before collision is equal to the total momentum after.

   c) 100.5 m s^{-1}

   d) Move the rifle closer to the putty.

2. a) 5 m s^{-1}

   b) 4 m s^{-1}

   c) 0.45 kg m s^{-1}

   d) 4.5 N

   e)

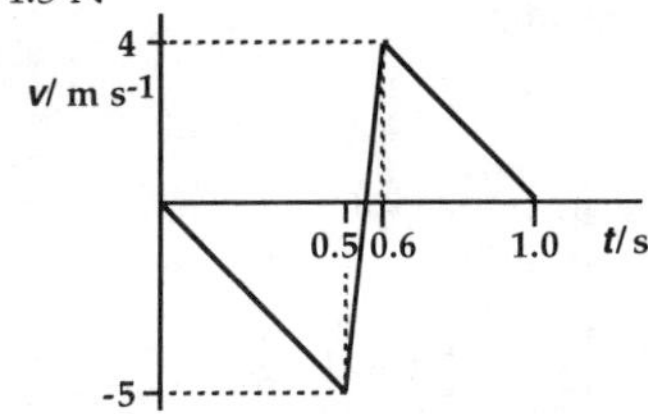

3. a) 34.6 m s^{-1}

   b) 20 m s^{-1}

   c) 34.6 m s^{-1}, 34.6 m s^{-1}, 34.6 m s^{-1};
   20 m s^{-1}, 10 m s^{-1}, 0 m s^{-1};
   40 m s^{-1}, 36.1 m s^{-1}, 34.6 m s^{-1};
   36.8 J, 30 J, 27.5 J

   d)

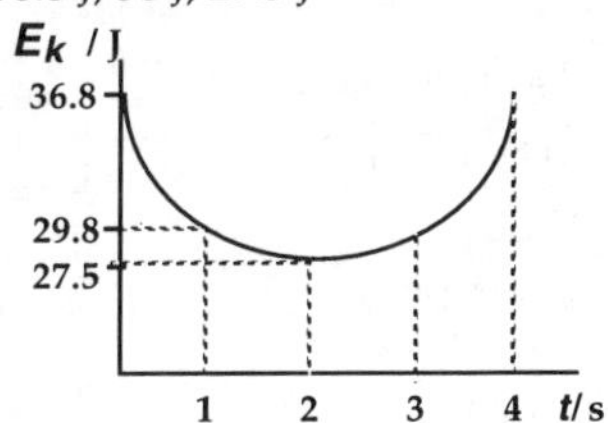

   e) 8.7 J

4. a) 0.027 m

   b)

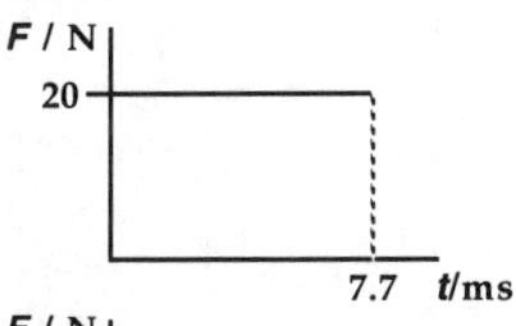

   c)

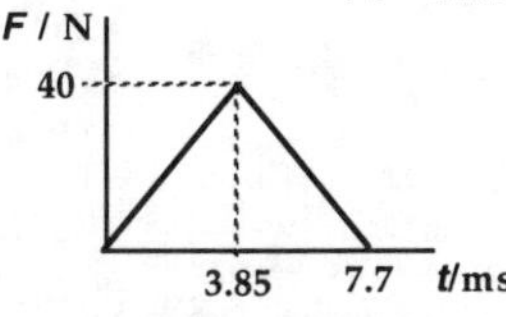

5. a) 0.505 s

   b) 1.48 m s^{-1}

   c) 0.1125 m

6. a) i) Calculate the quantity of energy from 0.7 kg of fuel. 1.96×10^5 J

   $E_k = \frac{1}{2}mv^2$

   Calculate v = 280 m s^2

   ii) All chemical energy goes to E_k.

   b) 840 m

   c) 128 m

7. a) 12.1 m s^{-1}

   b) 3.03 m s^{-1}

   c) No external forces acting.

   d) 11 025 J

   e) 3528 J

   f) 9.7×10^4 N

8. a)

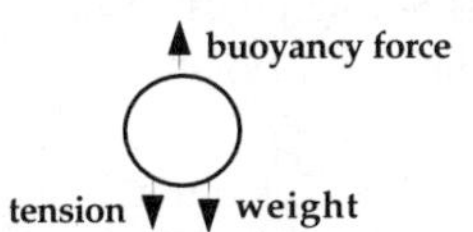

   b) i) 2.45 m s^{-2}

   ii) 24.5 m s^{-1}

   c) i) Initially moves upwards slowing down under gravity and then accelerates down.

   ii) 118 m

   d) Take reading on spring balance T,
   m = 1 kg W = 9.8 N
   acceleration $= \frac{T - 9.8}{1}$

PROPERTIES OF MATTER

Exercise 3.1 Density

1. 1000 cm^3

2. a) 1 x,
   b) 1 x,
   c) 1 x

3. a) 1000 x,
   b) 10 x,
   c) 0.001 x

4. 800 x
5. 4.4×10^{-6} m^3
6. 1.125 kg m^{-3}
7. 3.4 kg
8. 148 kg
9. 20.4 litres
10. Weigh flask + air = m_1
    Evacuate, reweigh flask = m_2
    Mass of air = $m_1 - m_2$
    Open under water,
    measure volume of water
    = volume of air removed.
    $\rho = \dfrac{m_1 - m_2}{V_{air}}$
11. density 1000 x, spacing 0.1 x
12. d, d, 10d
    d^3, d^3, 1000d^3
    $1/d^3$, $1/d^3$, $1/1000d^3$

Exercise 3.2 Pressure, Force and Area

1. 1960 Pa
2. 539 Pa
3. 1.2×10^5 N
4. 4×10^5 Pa
5. a) 7350 Pa
   b) 2.17×10^3 Pa
6. 7.9×10^4 Pa
7. a) 1.31×10^4 Pa
   b) ~1.47×10^6 Pa
8. 1.5×10^5 N
9. a) 1.68×10^5 Pa
   b) 3.55×10^5 Pa

Exercise 3.3 Boyle's Law

1. a) PV = constant when T and m constant
   b)

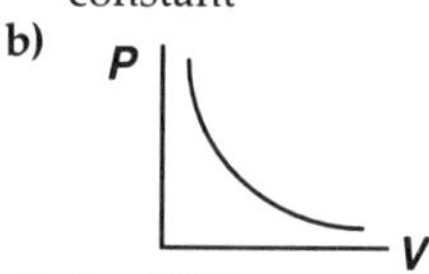

2. a) 4×10^5 Pa
   b) 5.625×10^5 Pa
3. a) 2100 litres
   b) i) 42 minutes
      ii) 41 min 24 s
4. a) 5 atmos.
   b) 3.3 atmos.
5. 400 cm^3
6.

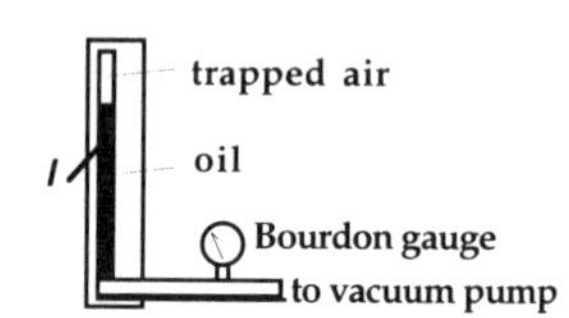

   Change P with foot pump, allow to come to room temperature.
   Measure $l \propto V$ and P. Repeat.
   Plot P against $1/V$.
   A straight line through the origin proves law.
7. 4.2 atmos.
8. 1.61×10^5 Pa
9. 60 hours
10. a) PV 1470, 1485, 1480, 1475
       $\therefore$ PV = 1478
    b) 296 kPa
    c) Longer tube, so larger volume of air not recorded, so larger error.

Exercise 3.4 Pressure Law

1. a) P/T = constant
   b) V and m constant
   c)

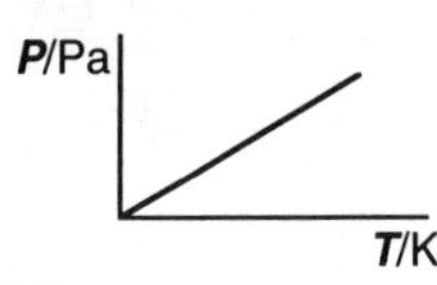

2. a) 77 K
   b) Lowest possible temperature at which all matter has stopped moving.

3. a) 33 °C
   b) 33 K

4. 3.53×10^5 Pa

5. a) 1.4×10^5 Pa
   b) 177 °C

6. a) P/T 0.35, 0.35, 0.35, 0.35, 0.35
   $\therefore P/T = 0.35$
   b) 68.2 K
   c) -278 °C
   d) Read P on meter. $T = P/0.35$

Exercise 3.5 Charles' Law

1. a) V/T = constant when P and m constant
   b)

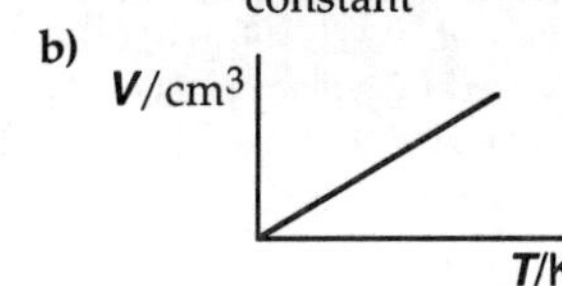

2. 327 °C

3. 5.89 litres

4. a)

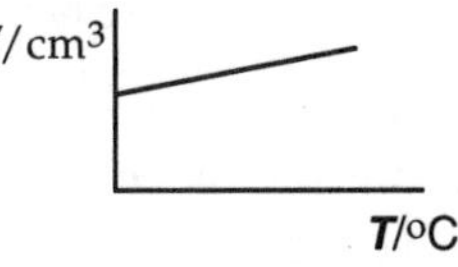

   b) Plot V against T in K.

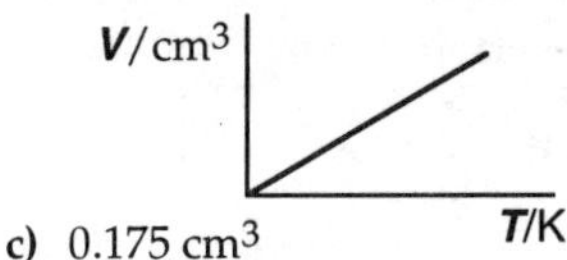

   c) 0.175 cm^3
   d) Place glass tube in freezing mixture read V of air, find T from graph.
   e) Changes in P affect V, Charles' Law only true at constant P.

6. -33 °C

7. 10.5 m^3

Exercise 3.6 General Gas Equation and the Kinetic Model

1. a) Collisions of air molecules with wall, each one exerts a force, total force divided by area = pressure.
   b) Gas particles have more E_k, so move faster.

2. a) $T\uparrow$ $\therefore E_k\uparrow$ $\therefore v_{particles}\uparrow$
   V constant, so distance between collisions constant.
   so more collisions. $\therefore P\uparrow$
   b) If P is constant, number of collisions is constant.
   $T\uparrow$ $\therefore E_k\uparrow$ $\therefore v_{particles}\uparrow$
   So distance between collisions must increase. $\therefore V\uparrow$
   c) T constant $\therefore E_k$ constant
   $\therefore v_{particles}$ constant if $V\downarrow$
   So distance between collisions must decrease,
   so more collisions. $\therefore P\uparrow$

3. a) $T\uparrow$ $\therefore E_k\uparrow$ $\therefore v_{particles}\uparrow$
   If V constant, distance between collisions is constant, so less collisions. $\therefore P\downarrow$
   b) $V\uparrow$, so distance between collisions increases.
   But T constant. $\therefore E_k$ constant
   $\therefore v$ particles constant, so less collisions $\therefore P\downarrow$

4. Assume V remains constant.
   $T\uparrow$ $\therefore E_k\uparrow$ $\therefore v_{particles}\uparrow$
   So more collisions. $\therefore P\uparrow$

5. 110 cm^3
6. 933 cm^3
7. 27 m^3
8. 77 °C
9. 1
10. 300 kPa
11. 6 x

Exercise 3.7 Pressure in Liquids

1. 2.33 m
2. 25 kPa
3. All the same
4. a) 1.034 x 10^5 Pa
   b) 2.57 cm
5. 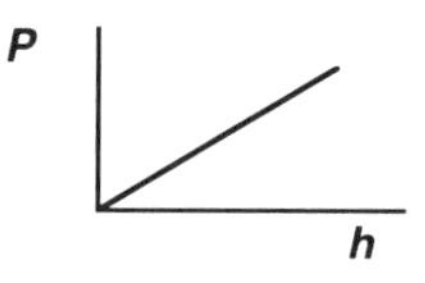

6. a) 1.11 x 10^5 Pa
   b) 11.5 cm
7. a) 11 cm
   b) 14.5 kPa
   c) 8.84 m
8. 71 cm

Exercise 3.8 Buoyancy and Flotation

1. P is proportional to depth
$P_{top} < P_{bottom}$
$F_{down} < F_{up}$
∴ net force up = buoyancy force helps to balance weight

2. 10.9 N

3. 2.43 m

4. a) 2.06×10^4 Pa
b) 2.35×10^4 Pa
c) 8.24×10^3 N
d) 9.41×10^3 N
e) 1.17×10^3 N
f) 1194 kg m^{-3}

5. Same weight as increased volume displaces same mass of air as contained in balloon.

6. Buoyancy force is the same, depends on volume but weight less as light polystyrene displaces water.

7. a) Upthrust ↓ as density ↓ since $P = \rho gh$
b) Depth ↑ since ρgh has to remain constant so that upthrust balances weight.

8. 209 N

9. 96.9 N

10. 13.3 N

11. $\rho_{iron} < \rho_{mercury}$ so weight of mercury displaced = weight of lead, when only part of volume of lead is submerged.

12. a) 5 N
b) i) Reading less since buoyancy force↑ as ρ ↑.
ii) No effect, as buoyancy force depends on difference between top and bottom depth and this is constant.

13. a) $F \propto$ depth, since A is constant and $P \propto$ depth.
b) Float tubes of same mass in liquids of varying ρ. Tube floats deepest in liquid of lowest ρ.
c) As ρ ↑ weight of liquid displaced ↑ ∴ buoyancy force ↑ ∴ $P = F/A$ ↑

14. a) 960 kg
b) 2272 N
c) Buoyancy force > weight
d) 1.67 m s^{-2}

15. a) i) 7.64×10^7 N
ii) 7.64×10^7 N
b) 77.2 m^3
c) $P = \rho gh$ ∴ ρ ↓ ∴ P ↓
∴ buoyancy force ↓
∴ acceleration ↓

Exercise 3.9 Mixed Problems

1. a) 6.67 atmospheres, T constant
b) 9.6 kg m^{-3}
c) i) 1.98 atmospheres
ii) 0.142 m^3
iii) 94.6 minutes

2. 1.1 kg m^{-3}

4. a) Q
b) S

5. 6 N

6. As handle pushed in V↑
∴ less distance between collisions
∴ more collisions ∴ P↑
E_k of handle↑
∴ E_k of particles↑ ∴ T ↑

7. a) 3000 N
b) Some air remains between wall and suction hook; wall will not be completely smooth.

8. a) Air diffuses out into the beaker ∴ water rises to balance decrease.
b) Air diffuses back in until P inside and outside pot the same, water falls to level in beaker.

9. a) AB Pressure Law
b) BC Boyle's law
c) -140 °C

RESISTORS IN CIRCUITS

Exercise 4.1 Resistors

1. 5 Ω

2. a) 2.3 Ω
   b) 3.23 Ω
   c) 2.77 Ω

3. a) C
   b) A

4. a) C
   b) A

5. a) 4 Ω 6Ω

   b)

   c)

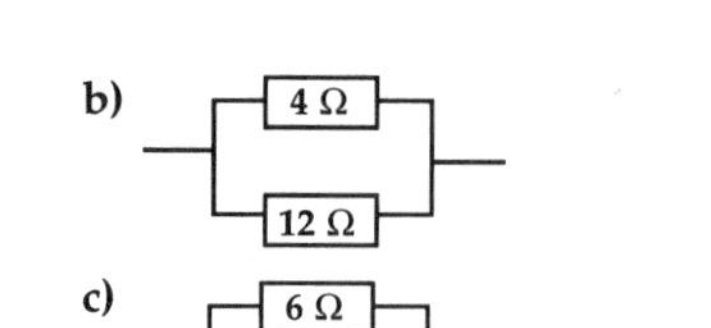

6. a) A↓
   b) V↑

7. a)

$$\frac{1}{R} = \frac{1}{R_1} + \frac{1}{R_2}$$

$$= \frac{1}{3} + \frac{1}{0.17}$$

$$= 0.33 + 5.88$$

$$= 6.21$$

$$R = \frac{1}{6.21} = 0.16\ \Omega \text{ per kilomete}$$

   b) Steel has physical strength.
   c) 25 600 W per km

8. 7.5 Ω

9. $R \times B = 12\,000$

Exercise 4.2 Potential Dividers

1. a) 3 V
   b) 4 V

2. 12 V

3. NP

4. a) 5 V
   b) 2.5 V

5. a) 1.5 V
   b) i) 0.56 V
      ii) $P = \frac{V^2}{R}$
      iii) 9.5 mW

6. 8 V

7. a) 60 V
   b) 48 V

8. a) 50 Ω
   b) 100 Ω
   c) 25 Ω

9. a)

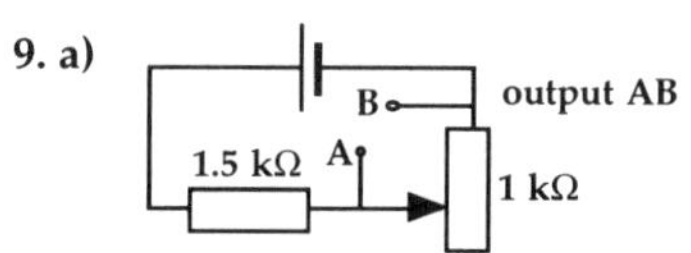

   b) i) Adding 1 kΩ in parallel.
      $\therefore R_{AB} \downarrow\ V_{AB} \downarrow$
      ii) 2.5 V

Exercise 4.3 Electrical Energy and Power

1. a) 18 W
b) 1.2 W

2. 64 J

3. 720 J

4. 250 V

5. a) 0.33 A
b) Nothing

6. 125 W

7. 15 W

8. a) i) 200 W
ii) 400 W
b) i) 400 W
ii) 800 W

9. 800 W

10. a) $P_B = 3P_A$
b) $P_A = 3P_B$

11. a) 2 A
b) 102 Ω
c) 3/20
d) Power dissipated in resistor.
e) Use a transformer.

12. a) 60 W
b) 2880 J
c) 15 120 J

13. 1000 J

Exercise 4.4 Circuits

1. 19 V

2. 10 Ω

3. 8 Ω

4. 3 Ω

5. 0.03 A

6. a) 2 A
b) 2.67 A

Exercise 4.5 E.m.f. and Internal Resistance

1. a) 4 A
b) 8 V
c) 4 V

2. a) 5 V
b) 3 A

3. a) i) 2 A
ii) 8 V
b) i) 3 A
ii) 6 V

4. 2.5 Ω

5. 1 A

6. 10 Ω

7. a) 12 V
b) 2 Ω
c) 6 A
d) 16 W

8. 3 Ω

9. a) The e.m.f. is the electrical potential energy supplied to each coulomb of charge which passes through the source.
b) i) 12 V
ii) 2Ω
c) As $I\uparrow$, lost volts $Ir\uparrow$
$\therefore$ voltage across external circuit(voltmeter)$\downarrow$

10. a) $V = IR_T \Rightarrow E = I(R + r)$

$$\frac{E}{I} = R + r$$

$$R = \frac{E}{I} - r$$

b) Compare to equation of straight line:
$y = mx + c$, $m = E$ and $c = -r$
c) 1 V, 0.25 Ω, 4 A

11. a) i) 5 A
ii) 1.6 V
b) 0.08 Ω
c) 11.1 W

Exercise 4.6 Wheatstone Bridge

1. 2 kΩ, 3.2 Ω, 10 Ω

2. 15 Ω

3. a) A -2 Ω
b) B +6 Ω, C +2 Ω, D -4 Ω

4. a) $T \downarrow R_{thermistor} \uparrow$
b) i) P ↑
ii) Q ↓

5. a) Zero
b)

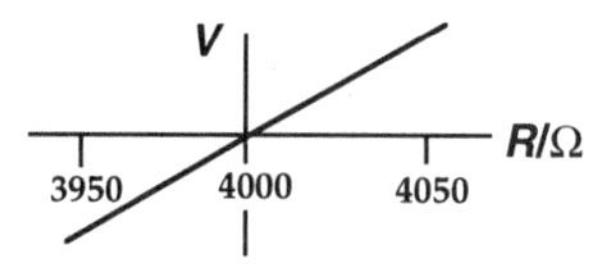

6. a) G reads zero => $\frac{R_P}{R_R} = \frac{R_Q}{R_S}$
b) Does not change, same effect on both branches *PQ* and *RS*.
c) Voltage can be high when bridge out of balance. Add resistor in series to **G** to protect **G** with a switch to bypass **G**. Close to balance.
d) 160 ± 10 Ω

7. a) 24 °C
b) 1520 Ω
c)

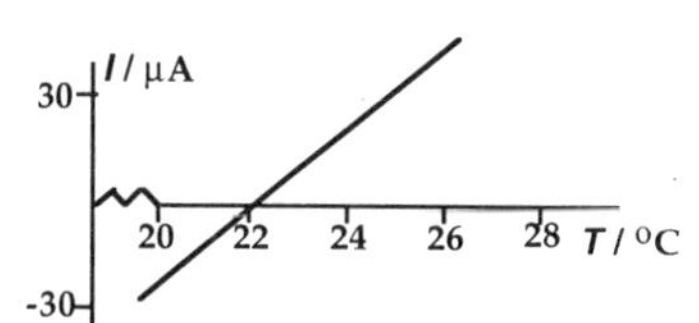

Exercise 4.7 Electric Fields

1. Particles experience a force and ∴ accelerate

2. a) **b)**

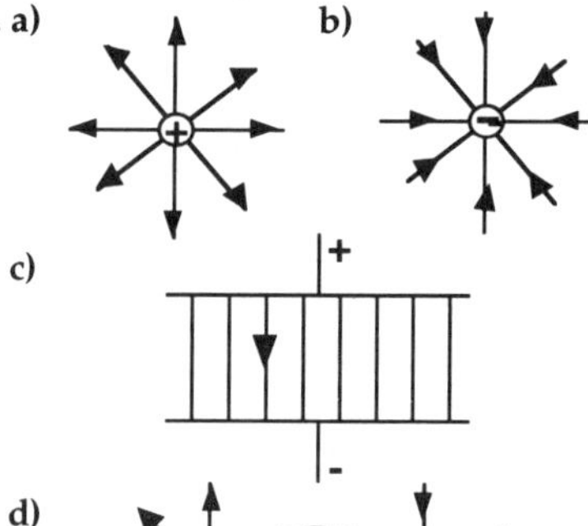

c)

d)

3. a) 12 J
b) 12 J

4. 8.38×10^7 m s^{-1}

5. 9.37×10^7 m s^{-1}

6. A p.d. of 20 V between the terminals means that 20 J of work is done in moving 1 C of charge between the terminals.

7. a) 4 times greater
b) 2 times greater

8. a)

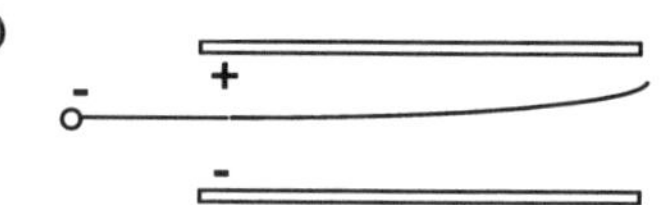

b) -2 units
c) -0.5 units

9. If 1 J of work is done moving 1 C of charge between 2 points the p.d. between points is 1 V.

10. 1000 J

11. a) $E = QV$
= charge on proton x voltage
b) 2.35×10^6 m s^{-1}

ANALOGUE ELECTRONICS

Exercise 5.1 Inverting Mode

1. a) 1250
b) 45.5

2. a) -0.2 V
b) -6 V
c) +12.9 mV
d) -0.425 V

3. a) +0.33 V
b) +1.14 V
c) -0.116 V
d) +1 V

4. a) 100 kΩ
b) 1 MΩ
c) 12 MΩ
d) 12.5 kΩ

5. a) 28.6 kΩ
b) 470 kΩ
c) 9.075 kΩ
d) 31.6 kΩ

Exercise 5.2 Inverting Mode Saturation

1. ±13.5 V

2. a) -1.5 V
b) -8.5 V
c) -13.5 V

3. a) -53.2 mV
b) -10.2 V
c) -10.2 V

4. a) 540 kΩ
b) 10.4 MΩ
c) 397 kΩ

5. a) 1.18 MΩ
b) 1.94 kΩ
c) 27.5 kΩ

Exercise 5.3 Inverting Mode, Analogue Inputs

1. a)

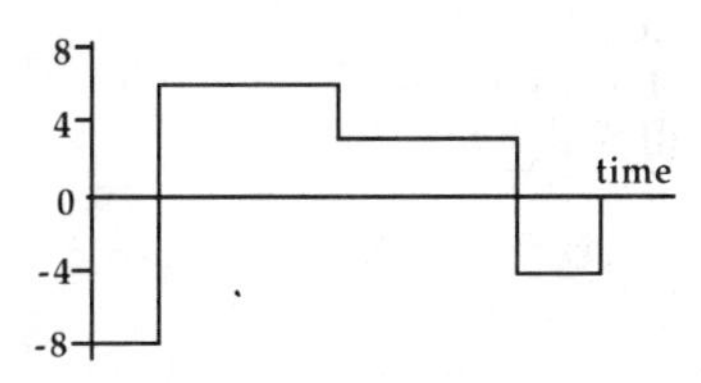

b)

voltage /V
16
saturation level
8
0
time
-8
-16

2. a)

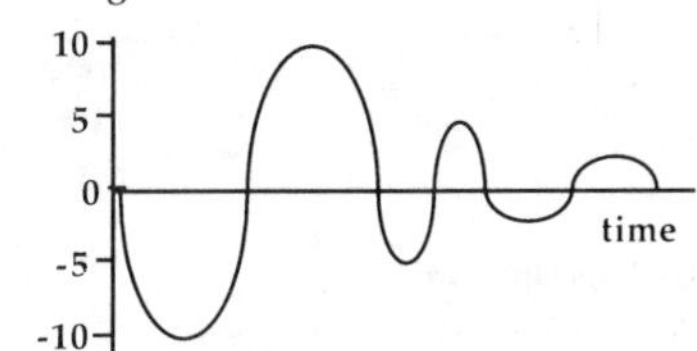

b)

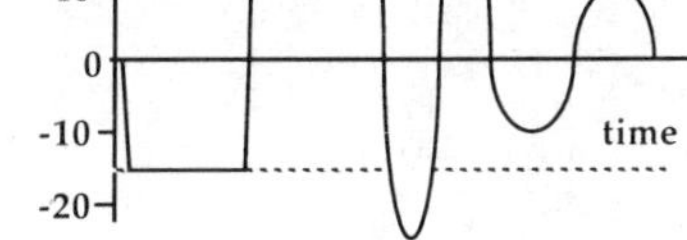

Exercise 5.4 Differential Mode

1. Amplifies difference between inputs.

2. a) 100 kΩ
   b) 264 kΩ

3. a) -2 V
   b) 8 V
   c) -4.75 V
   d) 87.4 mv
   e) 13.5 V

4. 603.2 mV

5. -0.193 V

6. -4.5 V to -13.5 V

Exercise 5.5 Differential Mode, Analogue Inputs

1. a)

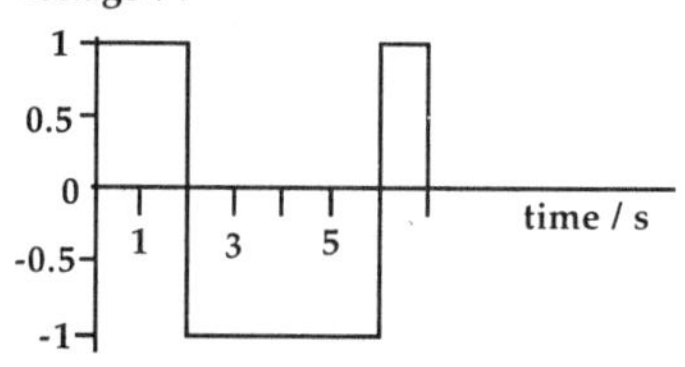

b)

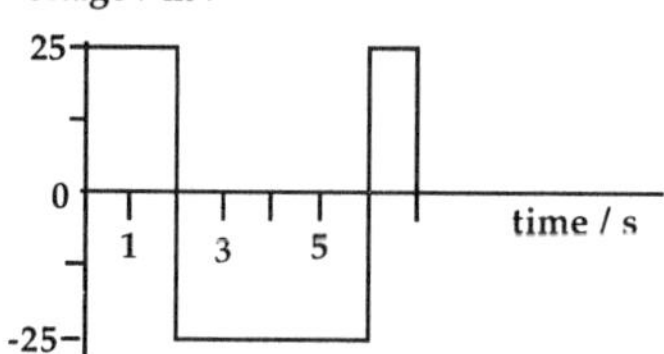

2. a)

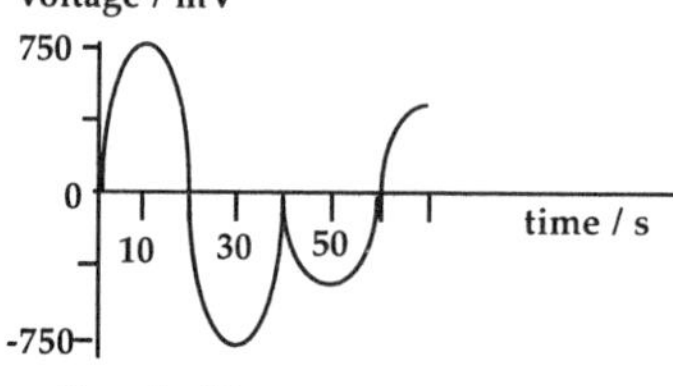

b)

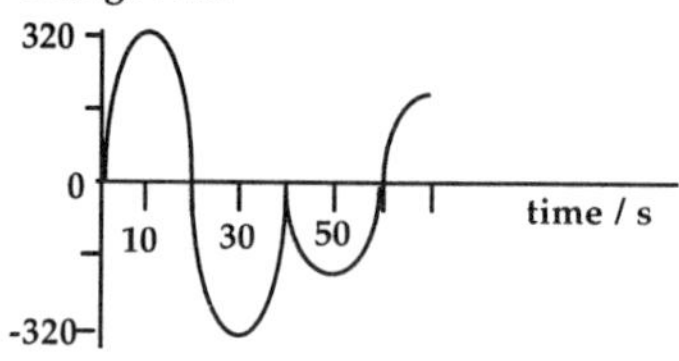

Exercise 5.6 Differential Mode, Monitoring Circuits

1. a) 5 V
   b) i) 8.93 V
      ii) -9.83 V
   c) i) 7.69 V
      ii) -6.73 V

2. a) 7.5 V
   b) i) 7.27 V
      ii) 3.48 V
   c) i) 7.66 V
      ii) -2.45 V

Exercise 5.7 Control Circuits

1. a) Differential mode
   b) 2.2 kΩ
   c) Light ↓ R_{LDR} ↑
      ∴ bridge unbalanced
      ∴ difference in inputs to op-amp
      Op-amp amplifies this voltage to switch on transistor.
   d) i) Transistor
      ii) Switches on and amplifies the current.
   e) 9.5 mV

2. a) 7.5 V
   b) i) 4.29 V
      ii) +13.5 V
   c) i) 12 V
      ii) -13.5 V
   d)

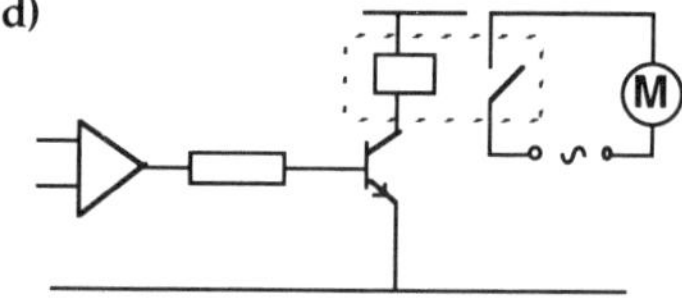

Exercise 5.8 Control Circuits

1. **a)** Input current is zero.
   No p.d. between inputs.
   b) Allows control of gain and gives stability with varying frequency.

2. **a)** Controls gain using feedback.
   b) -1.8 V
   c) Reverse 90 k and 10 k resistors.

3. **a)** -3 V
   b) ±7 V
   c) binary-decimal conversions
   d) Add more input resistors, each double the previous.

4. **a)** Op-amp in differential mode with original signal on one and the 50 Hz sine wave on the other.
   b) Op-amp amplifies the difference between the two inputs. The two 50 Hz signals cancel out leaving the difference in voltage to be amplified.
   c) 5 kΩ; 10 MΩ

5. **a)** 100 times
   b) 1 V
   c) 0.135 V

6. **a) i)** Light from window on solar cell causes output.
   ii) 0.219 V
   b) i) Differential mode
   ii) Variable resistor allows input to inverting input to be set equal to that at non-inverting input, so no output in normal light.
   iii) 0.407 V

7. **a)** R_V balances the Wheatstone bridge when no there is no load on the strain gauge.
   b)

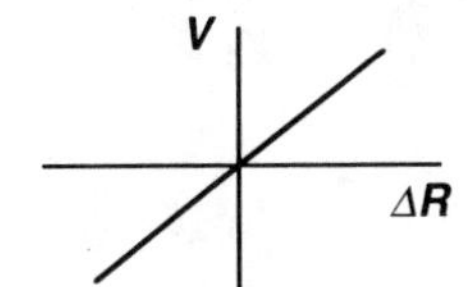

   c) 3 V

A.C CURRENT AND VOLTAGE

Exercise 6.1 Peak and r.m.s.

1. $I_{peak} = \sqrt{2}\, I_{r.m.s.}$
2. 325 V
3. 1.77 A
4. 2 A
5. 7.07 V
6. a) a.c. voltmeter
   b) oscilloscope
   c) 17 V
7. a) 10.6 V
   b) 4.5 mA
8. 8.5 V
9. 44.5 V
10. 21.2 V

Exercise 6.2 Frequency

1. a) 15 V
   b) 10.6 V
   c) 50 Hz
2. a) 50 V
   b) 35.4 V
   c) 66.7 Hz
3. a) 108 V cm^{-1}
   b) 5 ms cm^{-1}
4. a) 0.2 waves
   b) 1 wave
   c) 4 waves
   d) 10 waves

Exercise 6.3 Mixed Problems

1. 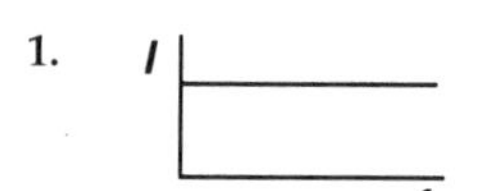

2. 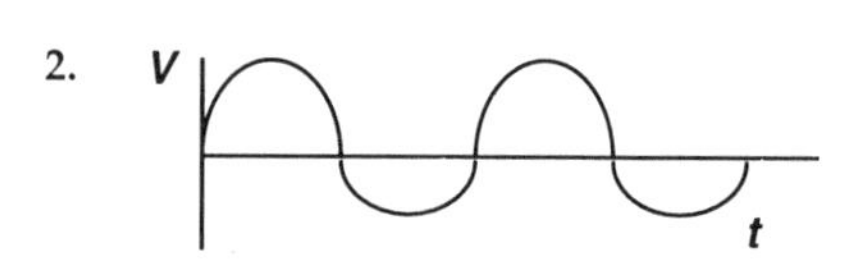

3. 4 V

4. 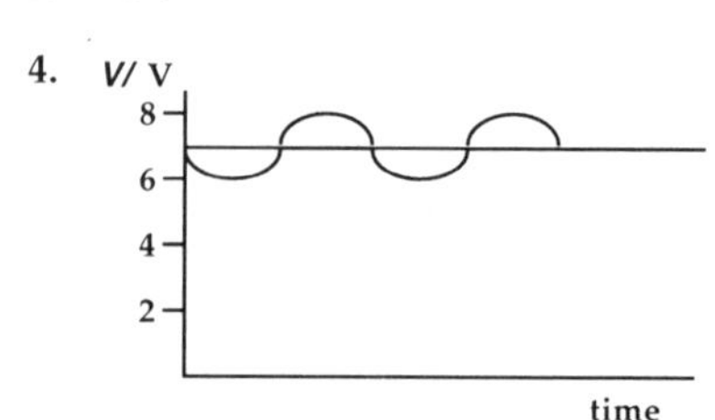

5. a) 30 V
   b) 500 Hz
   c) 16.25 vertical; 20 horizontal
6. a) **W** - 2 way switch
   X - variable resistor
   Y - light meter
   Z - oscilloscope
   b) Use the 2 way switch, connecting bulb first to the a.c. supply and then to the battery.
   c) Obtain the same reading on the light meter, keep it at the same distance from the bulb.
   d) Adjust the variable resistor to get same reading.
   e) Peak $V_{a.c.} = \sqrt{2}\, V_{d.c.}$

CAPACITORS

Exercise 7.1 Capacitance

1. a) Energy
   b) Capacitance

2. 40 μF

3. 4 μF means that the capacitance will store 4 μC for each volt across the plates.

4. D

5. 5 mF

6. a) 10 V
   b) 1 A

7. a) Reduce R
   b) 66.7 μF

8. a)

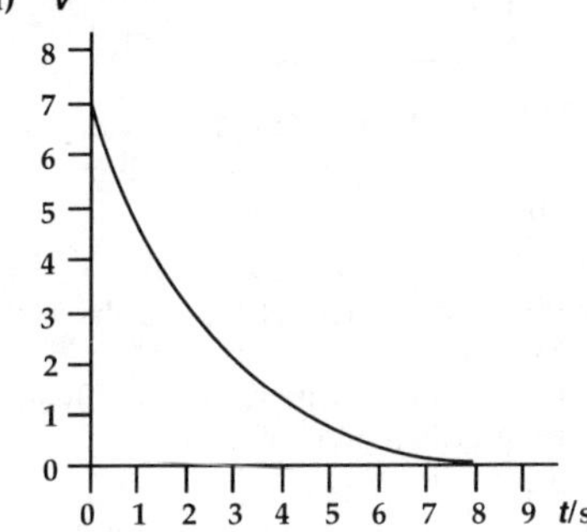

   b) 147 μA

9. a) 8 mC
   b) 160 μC
   c) 16 μF

10. a) 16 μF
    b) 17 ± 1 μF
    c) Increase the number of readings.

11. a) The ammeter gives a reading which then falls to zero.
    b) $I = 0$
    c) ± 1 mA
    d) None
    e) None

Exercise 7.2 Energy in Capacitors

1. $E = \frac{1}{2}QV = \frac{1}{2}CV^2 = \frac{1}{2}\frac{Q^2}{C}$

2. a) 2.2 nC
   b) 11 nJ

3. a) 32 mC
   b) 0.256 J
   c) Take care to connect the correct way round.
   d) The insulation breaks down and the capacitor is damaged.

4. a) 9 mC
   b) 54 mJ
   c) i) Less
      ii) Less

5. a) 3×10^{-10} F
   b) 15 μF
   c) 60 μJ
   d) 57.6 mJ

6. a) 0.0576 J
   b) The same since the same energy is stored in the capacitor.

7. a) i) 1.35 J
      ii) 846 W
   b) i) Large voltage could break down the insulation and damage the capacitor.
      ii) $I_{peak} = \sqrt{2}\, I_{r.m.s.}$
      $= \sqrt{2} \times 230$
      $= 325$ V
      ∴ too high to use

8. a) i) 1.28 mJ
      ii) 720 μJ
   b) 560 μJ
   c)

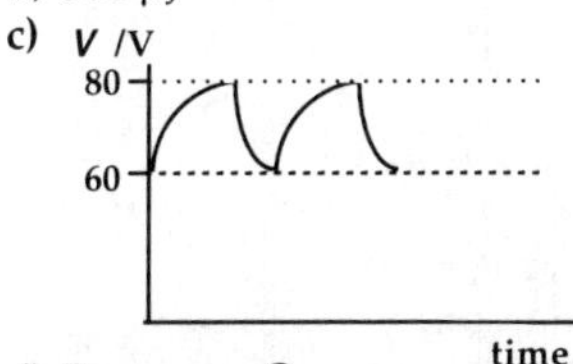

   d) Decrease C.
   e) Voltage across neon increases.
   f) Voltage across neon always > 80 V.

Exercise 7.3 Charge, Discharge Characteristics

1. a) 7.5 mA
   b) 4.23 mC
   c) 19 mJ
   d)

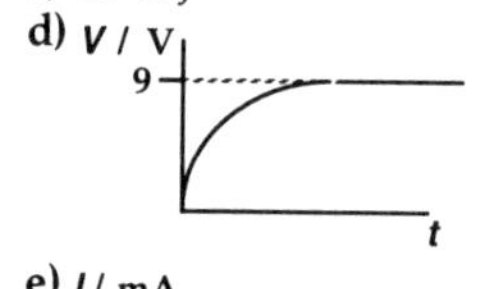

   e)

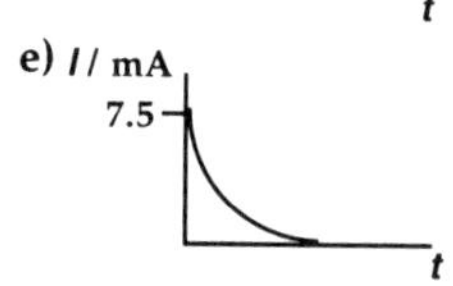

2. a)

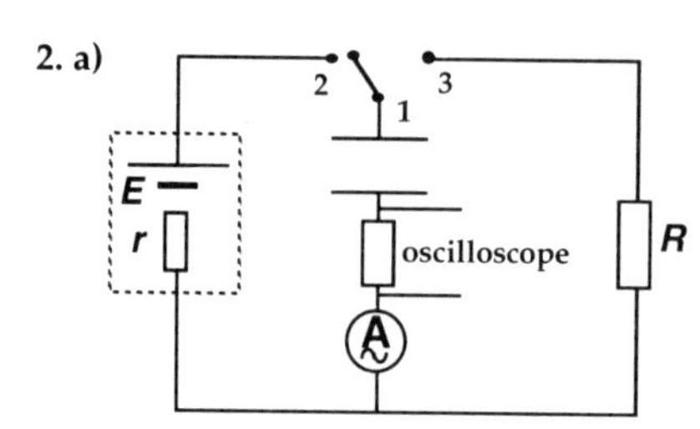

   b) Connect **1** to **2** to show charging and **1** to **3** to show discharging.

   c)

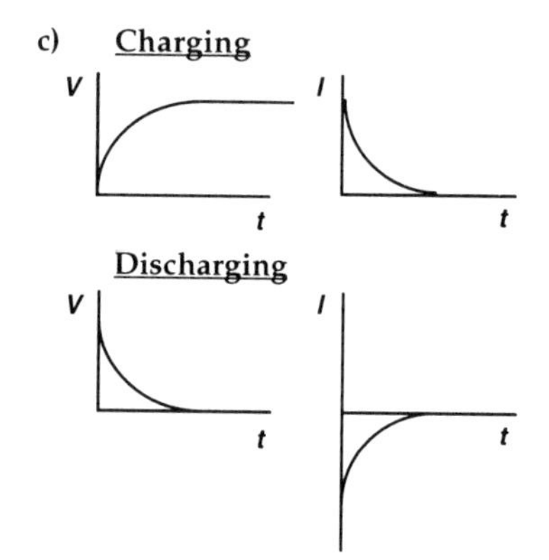

3. a) Discharge current
   b) Charging voltage
   c) Charging current
   d) Discharge voltage

4. a) 7 units, 5 s
   b) 3 units, 10 s

5. a)

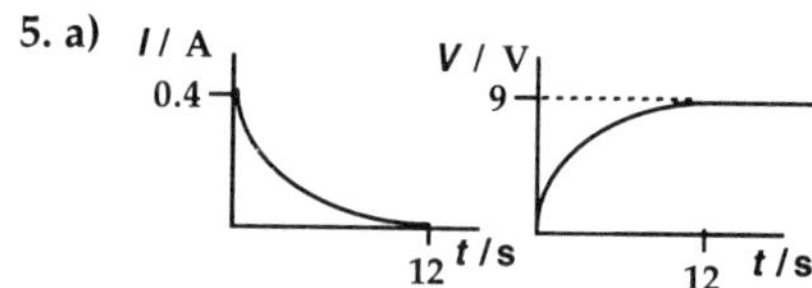

   b)

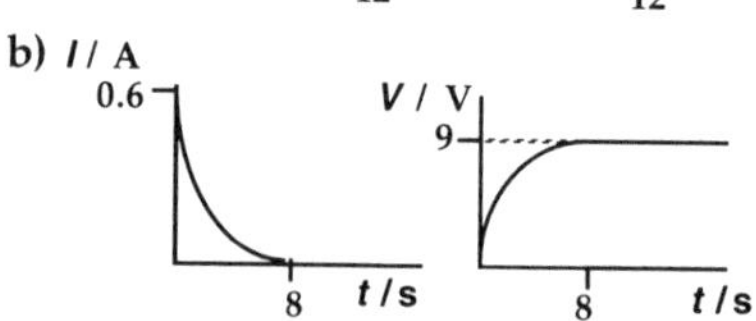

6. a) 6 V
   b) 7.5 mA
   c)

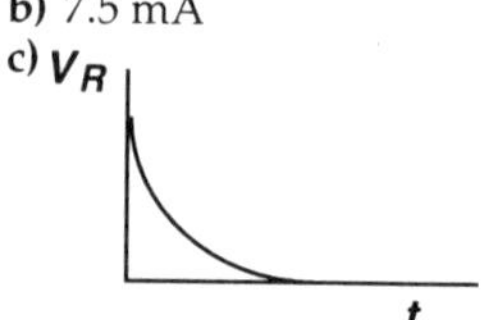

   d) Smaller than 800 Ω, the discharge current is more than the charging current.
   e) 0.18 J

7. a) Flicks in one direction and then falls to zero.
   b) Flicks in the opposite direction and then falls to zero.

8. a) 9 V
   b) Capacitor discharges though resistor.
   c)

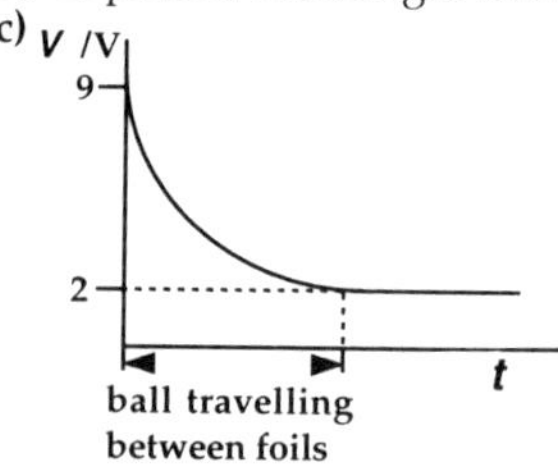

9. a) i) **AB** - Capacitor charging through resistor, neon out
   ii) **BC** - Capacitor discharging through neon, neon on
   iii) **CD** - neon goes out, capacitor starts charging again through the resistor
   b) Increase resistance (or capacitance)

Exercise 7.4 Capacitors in a.c.

1. a) I

   f

   b) I

   f

2. a) Brightness increases since the current increases as the current is directly proportional to the frequency.

   b) It goes out, since the capacitor blocks d.c.

3. 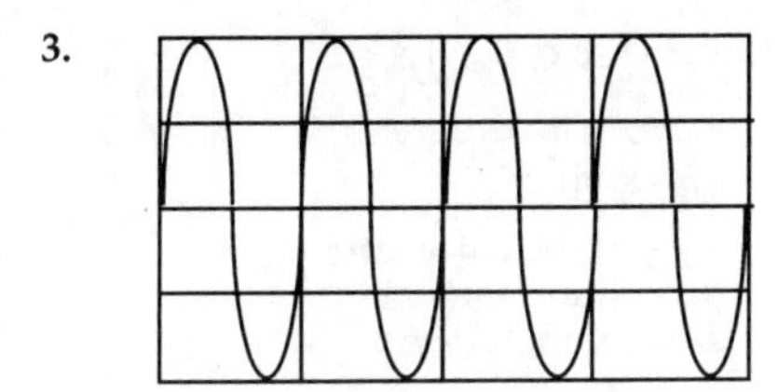

4. Low frequency rumble

OPTICS

Exercise 8.1 Waves

1. Interference
2. a) Smaller
   b) Same
   c) Smaller
3. a) 122 nm
   b) Smaller
4. a) Reflection, refraction, diffraction, interference
   b) Interference
   c)

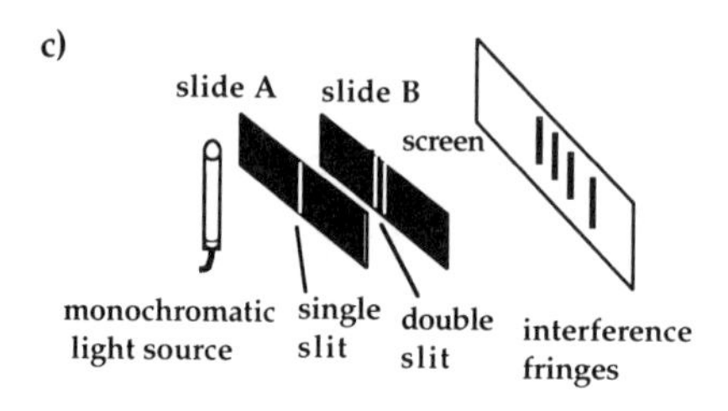

5. 0.02 m
6.

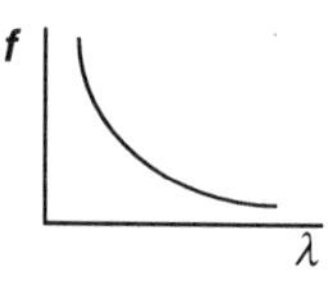

7. a) Radio, TV, microwaves, IR, visible, UV, X-rays, gamma rays.
   b) Violet, indigo, blue, green, yellow, orange, red
8. 22 m
9. $T = \frac{\lambda}{c}$
10. a) 0.60 m
    b) 15 m s^{-1}
    c) Halve, assuming the velocity is constant.

Exercise 8.2 Path Difference

1. a) **path difference** $= (n + \frac{1}{2})\lambda$
   where **n** is an integer
   b) **path difference** $= n\lambda$
2. $S_1X - S_2X = \lambda$ or 2λ
3. 800 nm
4. 4.5 cm
5. 0.16 m
6. 6 cm
7. maximum, $11.2 = 4 \times 2.8 \therefore n = 4$
8. 2y
9. 42.8 cm; 45.6 cm
10. Increase d, decrease L.
11. 4 cm, path difference = 6 cm = 1.5 λ
    $\therefore$ minimum
12. X - monochromatic
    Y - interference
    Z - amplitude
13. Wavelength is 0.34 m, so path difference is 0.5λ, so switching on decreases amplitude due to destructive interference.
14. a) Moving the reflector increases the path difference by 2 x distance moved.
    b) 3 cm
    c) i) 2.8 cm
       ii) 1.07×10^{10} Hz

Exercise 8.3 Prisms and Diffraction Gratings

1. a) One spectrum with red at the top, least deviated, violet at the bottom most deviated.
   b) Pairs of spectra with centre white, red furthest from centre most deviated and violet closest to the centre, least deviated.
2. a) 700 nm
   b) 540 nm
   c) 490 nm
3. a) $\lambda \downarrow$
   b) $f \uparrow$
   c) v constant
4. 8.6°
5. a) separation ↑
   b) separation ↓
   c) no change
   d) separation ↑
6. a) Diffraction grating
   b) Several pairs
7. a) 0.212 mm
   b) separation ↓
8. a) 400 nm
   b) 16 mm
9. a) i) Each colour has a different λ but $n\lambda = d \sin\theta$, so constructive interference occurs at different θ for each λ, so separation of colours.
      ii) 14.2°
      iii) 658 nm
   b) Only 1 spectrum not several pairs; red deviates least, not most.
10. a) i) 2.401 m
       ii) ± 0.005 m
    b) x
    c) 641 ± 6 nm
    d) Measure between the two 2nd order maxima and divide by 4 to find x more accurately.
11. a) i) Separation ↑
       ii) Separation ↑
       iii) Separation ↑
    b) i) For constructive interference, **path difference** $= n\lambda$ where **n** is an integer but red has a longer λ than blue light, so path difference has to increase.
       ii) At the centre path difference is zero, so **path difference** $= n\lambda$ is true for all values of λ, so all colours are present giving white.

Exercise 8.4 Refractive Index

1. a) Away
   b) Increases
   c) Stays the same
   d) Increases

2. 1.46

3. 1.8

4. a)

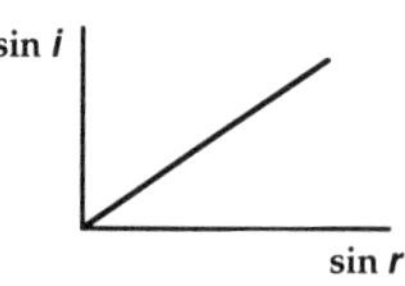

   b) Provided the ray enters at the centre of the straight edge, it will always meet the curved surface along a radius, i.e. at 90°, so no change of direction at the 2nd surface.

5. 1.35

6.

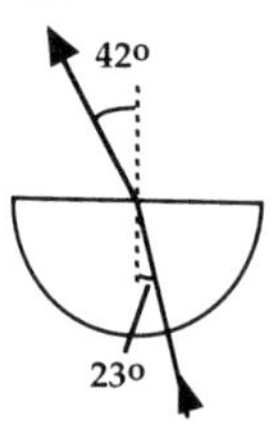

7. $n = \frac{\sin \theta_A}{\sin \theta_L} = \frac{\lambda_A}{\lambda_L} = \frac{v_A}{v_L}$

8. 9°

9.

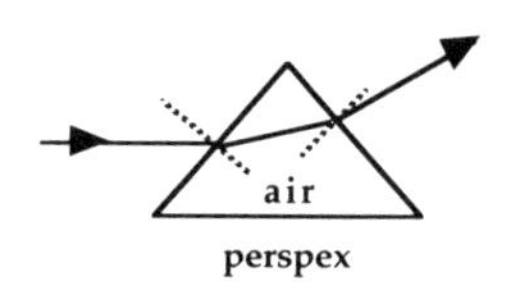

10. 2×10^8 m s^{-1}; 3.3×10^{-9} m; 6×10^{14} Hz

11. 0.2 m s^{-1}

12. a) Medium 1
    b) Medium 2

13. 2.14×10^8 m s^{-1}; 429 nm; 5×10^{14} Hz; 38.2°

14. No change in v or λ.

15. Medium 1 - water
    Medium 2 - air
    Medium 3 - glass

16. a) Plot a graph of sin **y** against sin **x**. Show that this is a straight line through the origin, so sin **y** = **n** sin **x** where **n** is a constant the refractive index.
    b) 1.51

17. a) P
    b) 0.53°
    c) red faster since $n = \frac{v_a}{v_m} \Rightarrow v_m = \frac{v_a}{n}$

18. a) 1.51
    b)

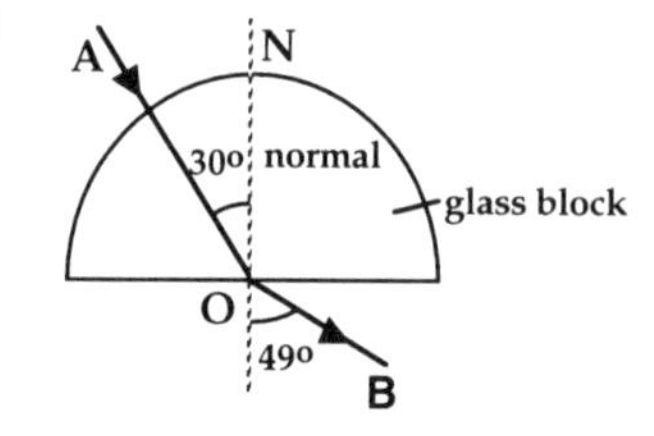

    c) 1.99×10^8 m s^{-1}

Exercise 8.5 Critical Angle

1. 36°

2. 24.4°

3. a) 1.5

   b) 41.8°

4. a) When light travels from the more dense to the less dense medium and meets a boundary at an angle > θ_c, all the light energy is reflected back into the more dense material.

   b)

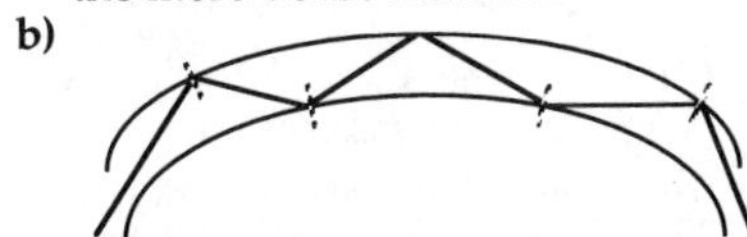

5. a) $n = \frac{v_1}{v_2}$

   b) 1.13

   c) 62.5°

   d) Medium 2

6. 31.1°

7.

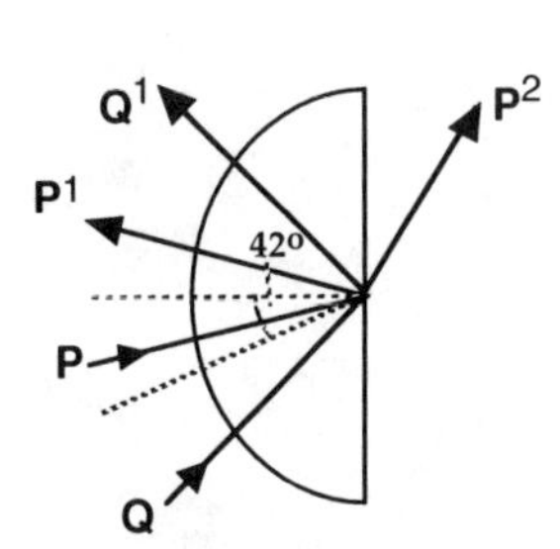

8. a)

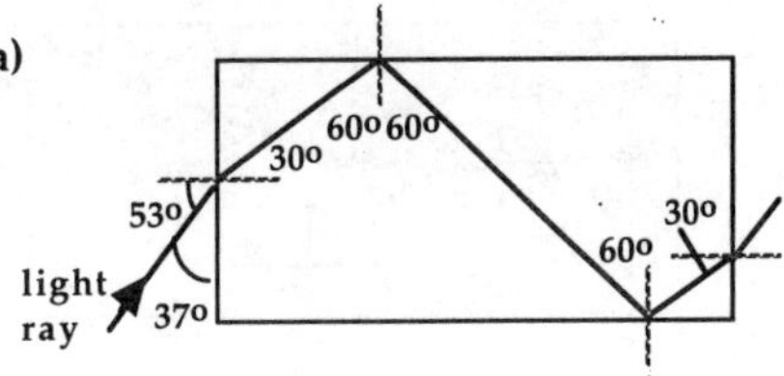

b)

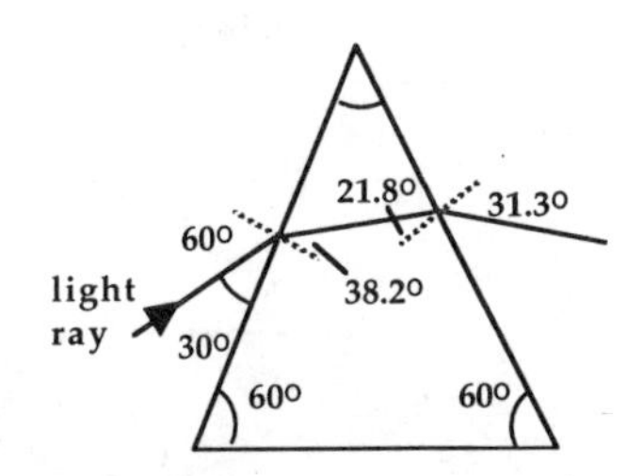

c)

c)

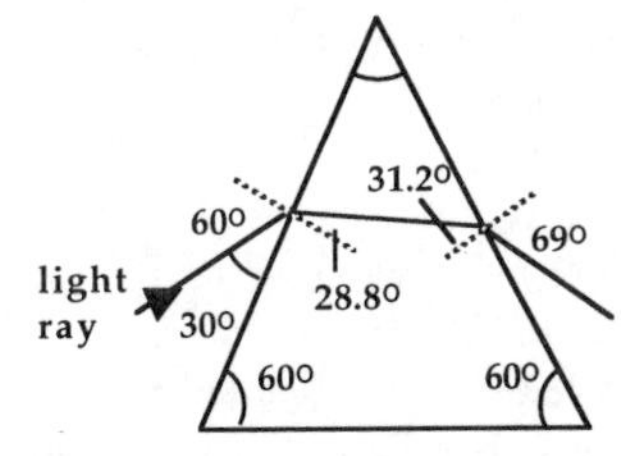

d)

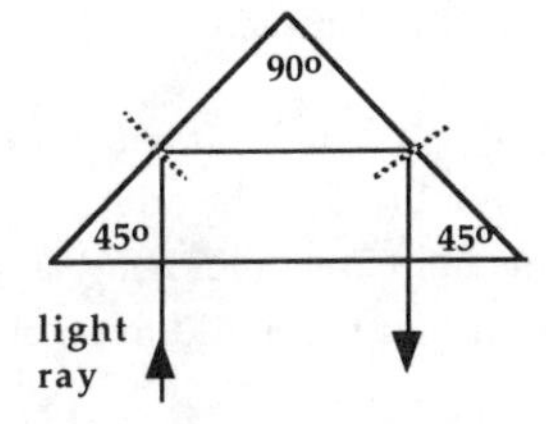

9. As $\theta_1 \uparrow$, **OP** becomes less bright, **OQ** becomes brighter. $\theta_2 \uparrow$ and θ_3 remains equal to θ_1, so also increasing. When $\theta_1 = \theta_c$ **OP** vanishes and light is totally internally reflected. **OQ** is as bright as the incident ray.

Exercise 8.6 Mixed Problems

1.

b) 23°

2. 3×10^8 m s^{-1}; 600 nm;
5×10^{14} Hz;
1.5×10^8 m s^{-1}; 300 nm;
5×10^{14} Hz;

3. a) 41.1°

b)

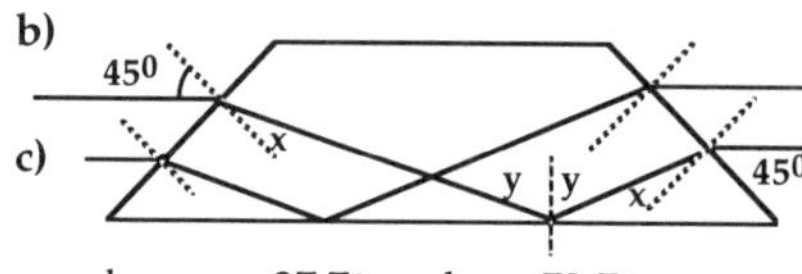

c)

where $\mathbf{x} = 27.7°$ and $\mathbf{y} = 72.7°$

d) The prism turns the image upside down.

4. a) 691 nm red

b) 487 nm blue

c) $\mathbf{x} > 48.6° \Rightarrow \mathbf{y} < \theta_c$ ∴ light refracts into the sample.
$\mathbf{x} < 48.6° \Rightarrow \mathbf{y} > \theta_c$ ∴ light totally internally reflected and does not reach the sample.

d) 60°

e) Colour **2** (blue) refracts more than colour **1** (red). The critical angle for blue light is smaller hence angle **x** will be less for the liquid to go bright.

5. a) 2×10^8 m s^{-1}

b) 41.8°

c)

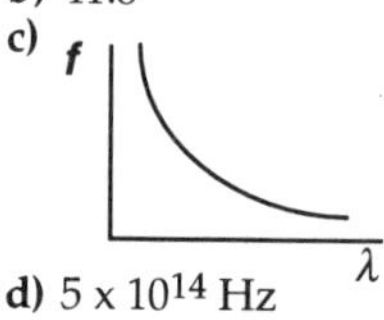

d) 5×10^{14} Hz

6. a)

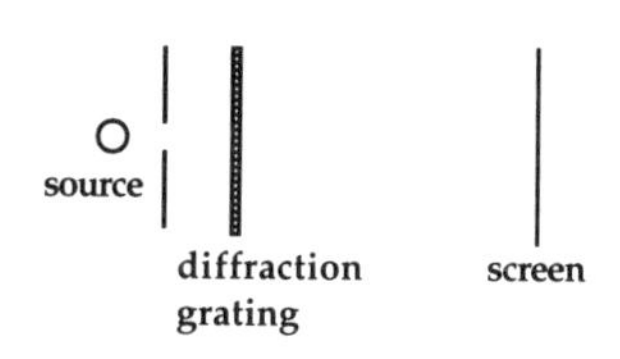

b) Maxima when $\mathbf{n}\lambda = \boldsymbol{d}\sin\theta$. This gives constructive interference because there is λ path difference between light from each adjacent slit in the grating.

c) 698 nm

d) For the new λ, $\theta\downarrow$ ∴ $\lambda\downarrow$

e) Move the screen back further by 27.9/24 times the original distance.

OPTOELECTRONICS

Exercise 9.1 Intensity

1. a) $I = \frac{P}{A}$

   P (Power - energy per second)
   A (Area - square metre)

   b) W m^{-2}

   c) $I = Nhf$

   N (number of photons)
   h (Plank's constant)
   f (frequency)

2. a) One quarter of the original intensity
   b) No change in λ

3. a) 11.1 W m^{-2}
   b) 400 W m^{-2}

4. 24 W m^{-2}

5. a) Move to 25 cm
   b) It acts as a point source.

6. D

7. a) 2.25 m^2
   b) 9 m^2

8. a)

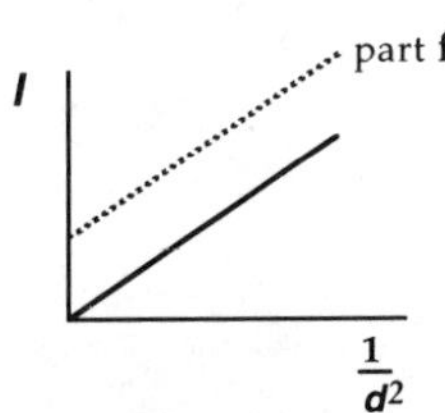

   b) $I = \frac{\text{constant}}{d^2}$

   c) 0.7 m
   d) Intensity
   e) ±0.07 m
   f) see part a)

Exercise 9.2 Photoelectric Effect

1. a) No discharge - positive electroscope
   b) Discharges
   c) No discharge - white light too low in frequency
   d) No discharge - white light too low in frequency

2.

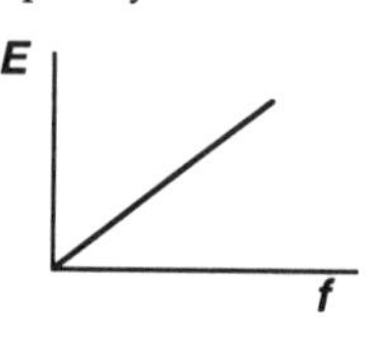

3. a) UV, so high f, so each photon has enough energy ($E = hf$) to allow one electron to escape. Dull light so few photons, so discharges slowly.
   b) White light, so f too small, so no photon has enough energy to release an electron. No electron absorbs more than one photon.
   c) More intense, so more photons, so discharges faster.

4. a) Blue light has higher f than red
   $\therefore E_{blue} > E_{red}$ but energy to escape is constant
   $\therefore$ blue light has more kinetic energy
   $E_k = hf - hf_o$
   b) Bright red light emits more photons, so releases more electrons than faint blue light.

5.

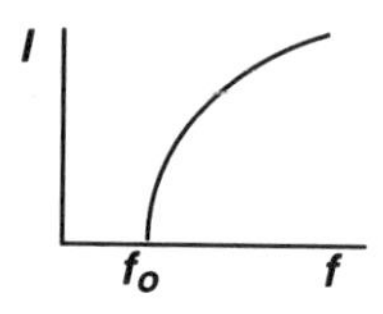

6. 4.9×10^{-19} J

7. 474 nm

8. a) True
   b) False
   c) True
   d) False
   e) False
   f) True

9. a) $E = hf$ Below f_o, E is not enough for any photon to eject an electron. No electron gains more than one photon.
   b) Light was considered a wave as shown by double slit experiment showing interference. Now had to be thought of as a particle, so wave-particle duality, i.e. photons as discrete bundles of waves.

10. 7.8×10^{-20} J

11. 5.2×10^{5} m s^{-1}

12. a) 5.58×10^{14} Hz
    b) 537 nm, yes
    c) 8.08×10^{-20} J
    d) 0.505 V

Exercise 9.3 Energy Levels

1. a) 6
b) i) E_3 -> ground
ii) E_3 -> E_2

2. a) E_4 -> E_0
b) 3.15×10^{15} Hz

3. a) i) B_3 -> B_2
ii) A_2 -> ground
b) If line is more intense, more electrons make the transition producing this line.

4. a) 10
b) 5.73×10^{14} Hz

5. $\lambda = \frac{hc}{E}$

6. a) i) 3 -> 2
ii) 6.56×10^{-7} m
b) Longest λ = E_5 -> E_1
= 9.5×10^{-8} m

Shortest λ = E_2 -> E_1
= 1.2×10^{-7} m

but visible spectrum is 4×10^{-7} m to 7×10^{-7} m $\therefore$ not in visible.

7. a) 2.67×10^{23}
b) 589 nm
c) -1.49×10^{-18} J

8. a) Potassium
b) 1.01×10^{15} Hz
c) $\Delta E = 5.9 \times 10^{-19}$ J or 3.7×10^{-19} J
d) i) Neither change is sufficient (below 7.1×10^{-19} J) to cause effect in copper.
ii) Both changes are sufficient (above 3.5×10^{-19} J) to cause effect in potassium.

Exercise 9.4 Spectra

1. Electrons in particular orbits round nucleus ∴ particular energy levels ∴ changes between energy levels give certain values of ΔE but $\Delta E = hf$ ∴ certain f ∴ only certain colours.

2. Gaseous sodium in Sun's outer layers absorb photons of these λ and electrons move to an excited level. Although same λ are emitted when electrons return to original level, they will not be in the original direction, so light missing these λ, so dark lines.

3. **Z** caused by more electrons making that particular transition.

4. Electrons only in certain well defined energy levels.

5. Mercury vapour gives a line spectrum, sharp lines of particular colours. Light bulb gives a continuous spectrum.

6. **a)** True
   b) False
   c) True
   d) True
   e) False

7. **a)** True
   b) False
   c) True

8. **a)** Use a diffraction grating to give spectra.
   i) Continuous spectrum - light bulb
   ii) Line emission spectrum - low pressure gas
   iii) Absorption spectrum - pass white light through low pressure gas.
   b) i) All colours red -> violet merging into each other.
   ii) Colour lines on black background.
   iii) Black lines on continuous spectrum.

9. **a)** A
   b) 2.92×10^{-19} J
   c) 2.84×10^{-19} J, no

10. **a)** Lower
    b) 3.38×10^{-19} J

11. **a)** In the sodium vapour lamp, electrons move down in energy level emitting photons of particular energy. When the photons pass through the vaporised sodium, the same element, they are absorbed by electrons in the lower level and the electrons move up in energy level. Since the photons do not reach the screen, there is a dark shadow.
    b) It has different energy levels, so the electrons do not absorb the photons emitted by sodium, so no dark shadow.

12. **a)** Spectra of tungsten in pairs on either side of central white line.
    b) Sodium vapour lamp gives different line spectra with strong yellow lines.

Exercise 9.5 Lasers

1. *Monochromatic* means single colour and therefore light of a single frequency (or wavelength).
   Coherent means waves with the same velocity, frequency and wavelength **and** with a constant phase relationship.

2. Light is coherent and concentrated in one direction.

3. a) "*Stimulated emission*" occurs when photons, having energy equal to the difference between two energy levels, trigger electrons to drop from a higher to a lower energy level, giving out radiation.
   b) Same frequency (wavelength); in the same direction; exactly in phase with the incident photon
   c) Have population inversion, mirrors reflect beam backwards and forwards along tube producing more photons.

4. a)

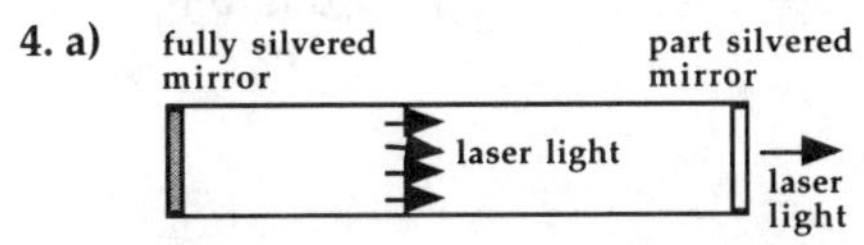

   b) Light, produced in the tube, reflects back and forwards between the two mirrors producing a high intensity of light in the tube. The partially reflecting mirror allows some light to pass through and this forms the laser beam.

5. a) 2.84×10^{13} Hz; 1.05×10^{-5} m
   b) Infra red

6. a) Although the power output is low, the light is concentrated into a narrow beam of small area which does not change with distance, so the intensity is high.
   Light bulbs emit light in all directions so the light is spread over a large area giving a low intensity.
   b) Surgery, e.g. repair of retina.

7. Maxima when $n\lambda = d \sin \theta$. This gives constructive interference because there is λ path difference between light from each adjacent slit in the grating.

8. a) 1.3×10^{11} W m^{-2}
   b) 2.45×10^{17}

9. a) 25 cm
   b) 457 J, yes

10. a) 5 s
    b) 9.85×10^{19}

Exercise 9.6 Semiconductors

1. a) Conductor - copper (any metal)
Semiconductor - silicon
Insulator - plastic or rubber

b) i) n-type - negative charge carriers, electrons, caused by impurities with 5 electrons in outer shell.

ii) Doping reduces resistance.

2. a) Electron
b) 5
c) Reduces resistance
d) No effect

3. a) Holes
b) Resistance of p-type is less.
c) Same - zero net charge

4. a) True
b) False
c) True

5. a) True
b) False
c) False

6. a)

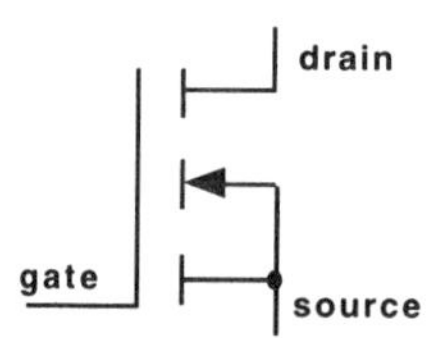

b)

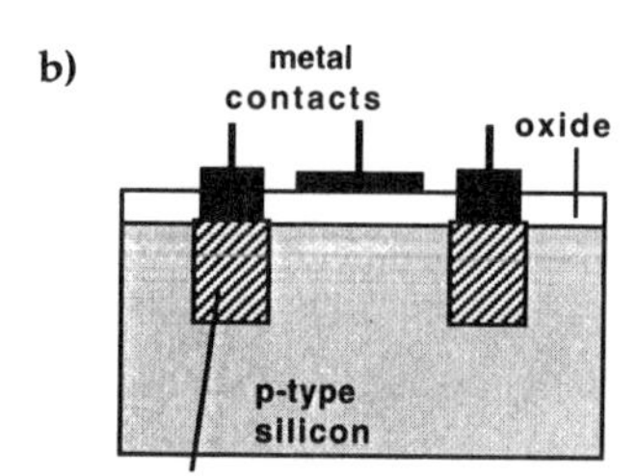

The contacts to the n-type areas are called the source and the drain. The contact over the oxide is called the gate.

c) The gate is made positive and this opens a channel for electrons between the source and the drain. The drain is always more positive than the source, attracting electrons through the channel.

d) Switch or amplifier

7. Switch or amplifier

8. Solar cell

9.

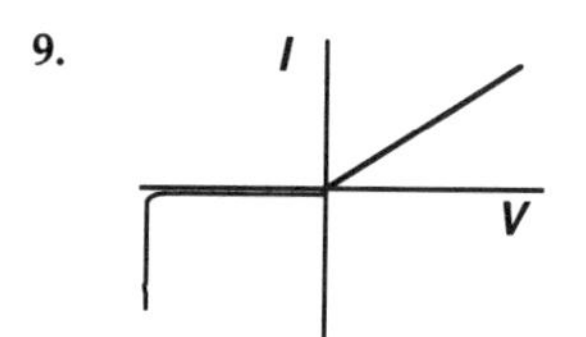

10. a) Dope with impurity atom with 1 less electron in outer shell.
b) Provides positive charge carriers, so reduces resistance.

11. a) Positive holes go to **B**.
b) 715 nm

12. a) Light produces electron-hole pairs at the p-n junction of the photodiode.
b) Photoconductive
c) 5.3 μA

13. a) 4.41 A
b) Light falls on p-n junction producing electron-hole pairs, giving a voltage.
c) 0.25 times less $I = \frac{\text{constant}}{d^2}$

RADIOACTIVITY

Exercise 10.1 Rutherford's Experiment

1. a) i) Alpha

ii) Very few particles at **Y**, slowly increasing. Most go straight through and then slowly decrease until very few at **X**. Scintillations on screen when particles hit.

iii) Since most go straight through, the mass must be concentrated. Few bounce back because of electrostatic repulsion.

b) Deflection less as electrostatic force less due to lower charge on nucleus.

2. a) Central nucleus containing positive protons and neutral neutrons. Negative electrons orbit nucleus. Same number of electrons and protons in the neutral atom.

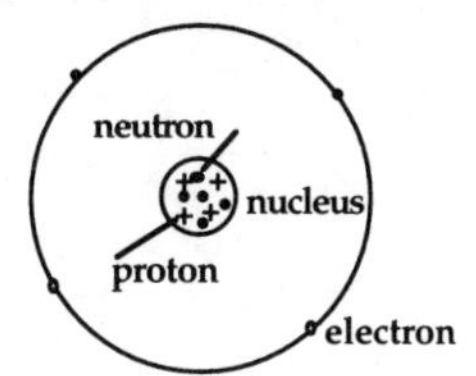

3. a) No
b) No
c) Yes
d) No
e) Yes
f) Yes
g) No
h) Yes
i) No

Exercise 10.2 Alpha, Beta and Gamma

1. 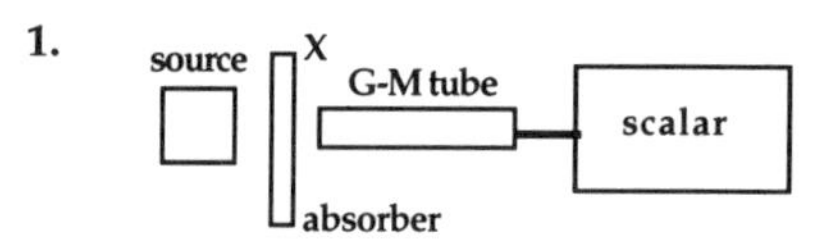

   Measure rate with source and no absorber. Put in paper at **X**, if rate falls α present. Put in few mm of aluminium at **X**, if rate falls β present. Put in few cm of lead, if rate falls γ present.

2. α, β

3. **a)** α,γ Paper absorbs α, γ will pass through thin lead but is absorbed by thick lead.
   b) Background radiation

4. **a)** 92
   b) 146
   c) 92

5. **a)** Isotopes have the same atomic number but different mass numbers.
   b) Atoms of relative atomic mass 35 are more abundant.

6. **a)** Ionisation occurs when a neutral atom loses or gains an electron.
   b) α

7. **a)** True
   b) False
   c) False
   d) False

8. **a)** α – helium nucleus $^{4}_{2}He$
   b) β – electron from the nucleus $^{0}_{-1}e$
   c) γ – high frequency short wavelength electromagnetic radiation

9. **a)** $^{234}_{92}U$
   b) $^{216}_{84}Po$
   c) $^{233}_{91}Pa$
   d) $^{238}_{94}Pu$

10. **x** - α; **y** - β; **z** - β

11. $^{224}_{88}Ra$

12. α; β; β

13. Z - 2; A - 4

14. **a)** -4
    b) -1

15. **X** - $^{208}_{81}Tl$; **Y** - $^{212}_{84}Po$

16. **a)** $^{139}_{57}La$
    b) γ rays

17. 212; 83; 212; 84

18. **a)** **x** - α; **y** - β; **z** - α
    b) 5 α and 3 β
    c) Could also be emitting γ rays, since these do not affect the isotope.

Exercise 10.3 Absorbed Dose and Dose Equivalent

1. 0.03 kg
2. 0.5 mGy
3. a) 2 mSv y^{-1}
   b) Cosmic rays from sun; radon gas; rocks and soils; living things
4. a) Energy absorbed per unit mass
   b) Photographic badge with film covered by varying thicknesses of absorbing material.
   c) $H = DQ$
   d) 5 days
5. a) Fast neutrons
   b) 12.5 hours
   c) 20 μJ
6. a) i) α, β
   ii) 49.2 μSv
   iii) No, less than 5 mSv
   b) Extra shielding; sit further away
   c) Energy of radiation; distance from source; type of radiation and shielding
7. 3.6 mSv per year, no
8. a) 1.8×10^{10}
   b) 3.125 hours
9. a) 10 kBq
   b) 6 nGy
   c) 10.2 nSv

Exercise 10.4 Half Life

1. 8 mins
2. 3.20 pm
3. 20 c.p.s.
4. Nothing
5. a) γ
   b) Few hours
6. 8 mins
7. 100 c.p.m.
8. 17 c.p.s.
9. a)

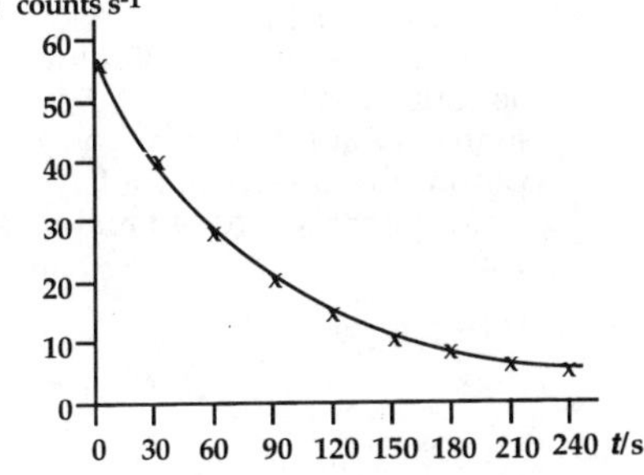

   b) 6 c.p.s.
   c) ~54 s
   d) After 1 half life the background count rate would be about the same as the rate due to the source. After this, the rate due to source becomes negligible, so inaccurate.
10. a)

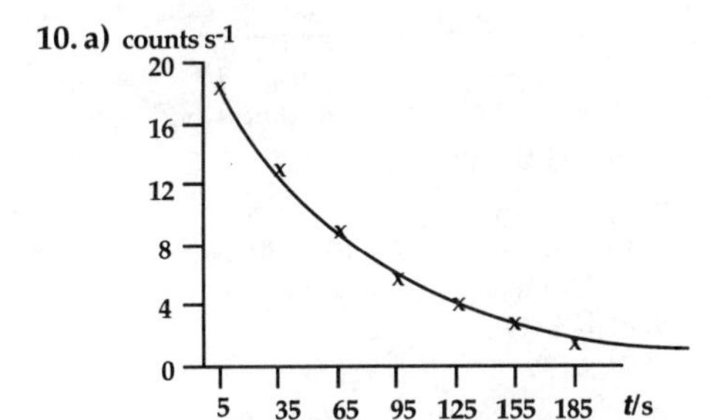

   b) 60 s; use graph to find time for initial activity to half.
11. ~3.1×10^{9} years

Exercise 10.5 Half-value Thickness

1.

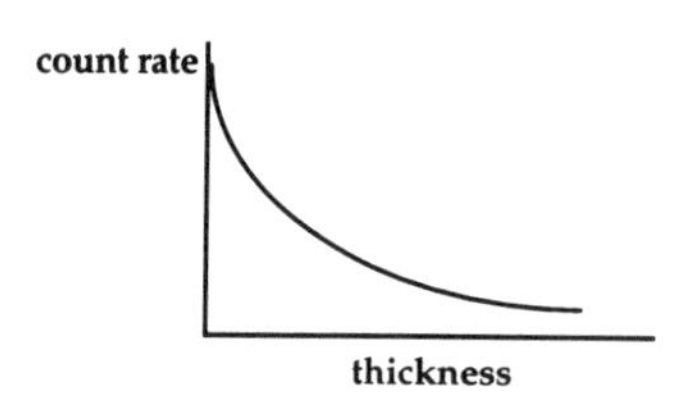

2.

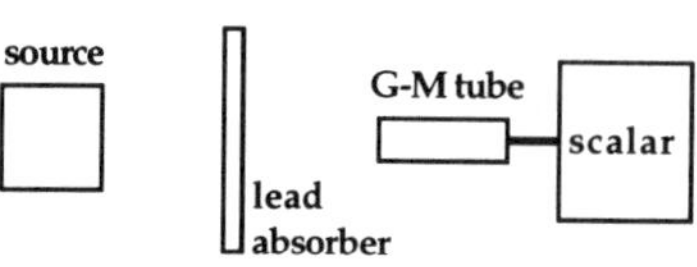

Keep the G-M tube at a constant distance from the source.
Measure the rate over 30 s for:
no source,
source but no lead,
various thicknesses of lead
Correct for background and plot graph.

3. 1/32

4. $\frac{1}{128}$ *I*

5. 30 c.p.s.

6. 24 mm

7. a)

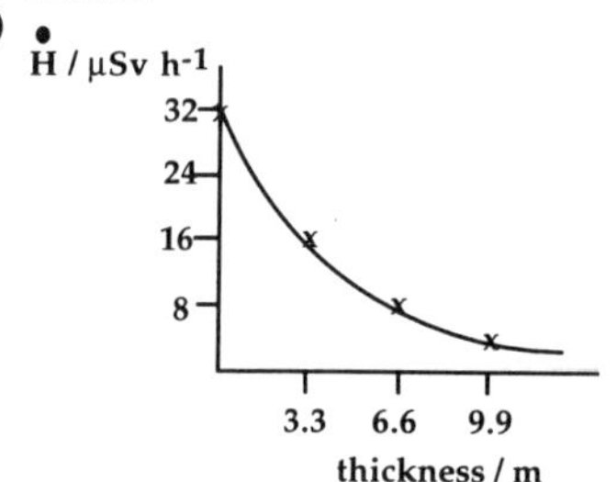

b) ~11.9 mm

8. a) 36 mm

b) DQ neutron = 2 x 10 = 20 Sv
DQ shielded gamma = 5 x 1 = 5 Sv

9. a) 7.2 cm

b) 8 m

10. With shield 1.1×10^{-5} Gy,
without shield 2.2×10^{-5} Gy,
so use shield

11. a) 6×10^5

b) i) 8 μGy
ii) 3 m 50 cm

Exercise 10.6 Fission and Fusion

1. a) Two small nuclei combine to form one larger nucleus.

b) Not all the mass of the small nuclei is needed for the large one. $E = mc^2$, so extra mass gives energy.

2. a) Induced fission reaction

b) Mass of $^{235}_{92}U$ plus neutron is greater than the mass of the daughter products.
$E = mc^2$, so extra mass gives energy.

3. **P** = 90; **Q** = 56

4. a) 92 protons and 235 - 92 = 143 neutrons in uranium nucleus.

b) **x** = 134; **y** = 40

5. $^{9}_{4}Be + ^{4}_{2}He \rightarrow ^{12}_{6}C + ^{1}_{0}n$

6. a) hydrogen $^{1}_{1}H$

deuterium $^{2}_{1}H$

tritium $^{3}_{1}H$

b) i) $^{2}_{1}H + ^{3}_{1}H \rightarrow ^{4}_{2}He + ^{1}_{0}n$
ii) Fusion
iii) In the Sun (stars)

7. $^{13}_{7}N \rightarrow ^{13}_{6}C + ^{0}_{-1}e$
carbon 13 is formed.

Exercise 10.7 $E = mc^2$

1. a) Fusion reaction
b) 9×10^{-13} J

2. 2.61×10^{-12} J

3. a) $^{239}_{94}Pu + ^{1}_{0}n \rightarrow$

$$^{108}_{46}Pd + ^{129}_{54}Xe + 3^{1}_{0}n + 6^{0}_{-1}e$$

b) 4.05×10^{-11} J

4. a) $^{1}_{1}H$ - 1 proton and 1 neutron

$^{2}_{1}H$ - 1 proton and 2 neutron

$^{3}_{1}H$ - 1 proton and 0 neutron

b) i) 7×10^{-30} kg
ii) 6.3×10^{-13} J
c) 5.7×10^{38} fusions every second and in each one 7×10^{-30} kg are changed into energy.

5. a) Induced fission
b) Spontaneous fission
c) **a** = 56; **b** = 1
d) i) **A** = 3×10^{-28} kg; **B** = 4×10^{-28} kg
ii) Reaction **B** releases more energy as more mass is 'lost', changed into energy.

6. a) **x** = 2; **y** = 7
b) 2.925×10^{-11} J
c) 349 MW

MISCELLANEOUS

Exercise 11.1 Uncertainties

1. a) 1.74
b) ± 0.01 m

2. a) 5.5 ±0.25 V
b) Half the smallest scale division

3. 15 ± 0.08 Ω

4. a) 600 J kg^{-1} $^oC^{-1}$
b) Rise in temperature - 10%
c) ±60 J kg^{-1} $^oC^{-1}$

5. a) i) 8.85
ii) ± 0.07 ms
b) 339 ± 2 m s^{-1}
c) Make more measurements of time.

Exercise 11.2 Mixed Problems

1. 61 %

2. a) $y \propto x^2$
b) $y \propto \sqrt{x}$
c) $y = mx - c$

3. a) i)

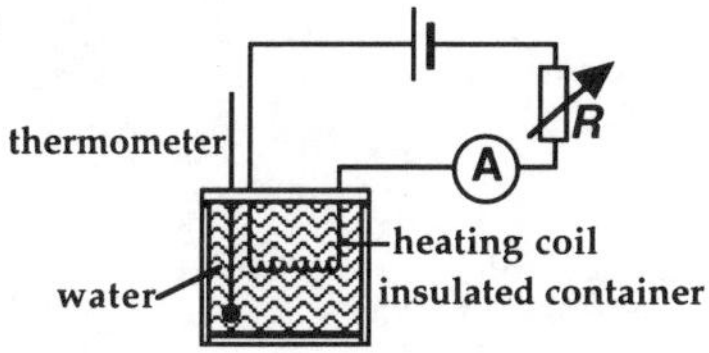

thermometer
heating coil
water
insulated container
A
R

ii) Use variable resistor ***R*** to vary current. Run for time ***t*** (say 30 s) and measure ΔT.
Measure m_{water}. Use c_{water}.
$E_h = cm\Delta T \therefore P = E_h / t$.

iii) Lid on container; use insulation to minimise heat loss

b) i) Rate of heat produced
= constant x I^2

ii) 4 Ω

4. N s = kg m s^{-2} x s = kg m s^{-1}

5. V m^{-1} = J C^{-1} m^{-1} = Nm C^{-1} m^{-1}
= N C^{-1}

6. Watts

7. kg^{-1} m^3 s^{-2}

8. a) Diffraction
b) $\lambda_{uv} < \lambda_{visible}$
c) Electrons act as waves.

9.

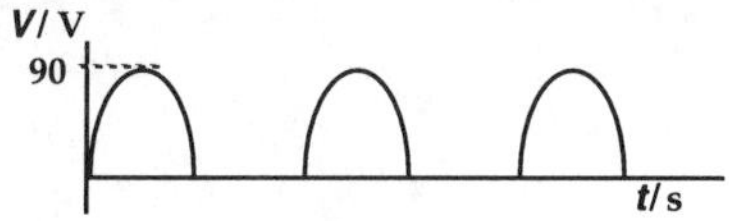

V / V
90
t / s

10. 1.48 A

11. a) 141 nm
b) 2.13 x 10^{15} Hz
c) 1.41 x 10^{-18} J; yes
d) 400 nm violet
700 nm red
e) λ_o = 710 nm ∴ works in visible

12. a) p-n junction near the surface.
b) Photons of light on junction form electron-hole pairs, producing a voltage.
c) 0.067 m^2
d) 66.7 J
e) 0.79°

13. 6.3 ± 0.1 kW

14. a) 18.7 J
b) 153 W
c) 97 W
d) 38.8%
e) Energy heating up filament and glass, decreases efficiency; energy lost by metal block to air, if included would increase efficiency; actual power of bulb may not be 250 W, could increase or decrease efficiency.
f) Side facing bulb black as black is the best absorber; other sides not black as black is also best radiator.

15. a) i) 0.12 °C
ii) Some energy lost as kinetic energy.
b) i) Large change in length of column of mercury for small ΔT.
ii) Make tube containing mercury very thin and long.